The Enigma
That Was
Thomas William Cowan

by Robert J Hawker

Northern Bee Books

The Enigma That Was Thomas William Cowan

© Robert J Hawker

ISBN 978-1-904846-75-8

Published by Northern Bee Books, 2011
Scout Bottom Farm
Mytholmroyd
Hebden Bridge HX7 5JS (UK)

Design and artwork by D&P Design and Print
Printed by Lightning Source UK

The Enigma

That Was

Thomas William Cowan

by Robert J Hawker

Contents

Preface

Sixteen years ago, when reading W. Herrod-Hempsall's massive two volume *Bee-keeping New and Old,* and putting a few notes on file, I typed in the margin 'is this true'? Later I repeated the question five more times. Each query concerned an entry on Thomas William Cowan. Thus started my retirement 'default' activity, the result of which manifests itself in the following pages.

As time passed I became 'drawn into' Cowan's World. It was a puzzle, and I enjoy such challenges. I have read every word in the *British Bee Journal* and *Beekeepers Record,* from their commencement until 1926 and selected pieces thereafter; all the beekeeping entries in *The Journal of Horticulture* for the same period; *Beemaster;* the early years of *the English Mechanic and World of Science; The Irish Bee Journal;* searched on line *The American Bee Journal* and *Gleanings;* a large number of books, all those authored by Cowan of course, and every edition of each. Those I found helpful are listed in an index.

This resulted in many files containing notes made as I read, which were then linked into other files, in theory making every word on any subject easily accessible at the press of a few buttons, as I sit at my desk. Needless to say modern technology has had a large part to play. I hope that I have put the pieces together and produced a readable volume, but in doing so, some of the minute detail has been omitted – my initial attempt was nearly three times as long!

However, the great joy of such an activity, I discovered, was not the attempted mastery of the technology or the puzzle solving, but interacting with many individuals previously unknown to me, with diverse interests, who were prepared to give their time and energy dealing with my queries. I hope that I have acknowledged their assistance adequately.

If this preface had been written in my subject's period, I would be claiming that there was a large demand for such a work and I had received numerous requests to produce it; not a book like it on the market; produced with no expense spared; etc. It is true that one or two friends have suggested that it would be a good idea to make my research public, and that there is no such volume on the market, although there have been numerous magazine sketches, some badly researched. It is also a fact that it has been subject to considerable expenditure, but it is a rare hobby that is free - excepting bee-keeping of course, where we generally cover our costs, the pain and pleasure costing nothing. However, I do not deceive myself that the great British public are likely to be queuing up to read it.

I have attempted to discover the real Thomas William Cowan, a difficult task given his secretive nature and no living direct descendents. I traced two of his relatives but both declined to assist, believing that they had no information worth supplying. Two very distant relatives helped with the family history and properties but neither knew anything relating directly to Thomas and family. Perhaps the appearance of this work will bring others forward.

There are many instances where independent corroboration of a 'fact' has not been possible, and I also acknowledge that it is not possible to absolutely prove a negative. The narrative contains examples of both, as it does the occasional attempt to interject a lighter note for which I make no apology.

Dimensions are given in those used at the time – Imperial, and the same applies to money. Reference to the internet or set of tables will provide the necessary conversion if required.

Some people have helped and encouraged me more than others, probably exhibiting either great foolishness or misplaced confidence; and it would be very remiss of me not to specifically acknowledge them. David Charles has been unselfish in his encouragement, support, and supply of information. I thank him particularly for several photographs and the relevant unpublished parts of the Henry Michell diaries, which provided the key that unlocked much of T. W. Cowan's early life. The late Len Davie greatly assisted by searching some of the more obscure publications for me, and I unashamedly used him for his expert information on bee diseases that provided the backdrop for the chapter on legislation, although not included in detail. I thank Una Robertson of the Scottish Bee-keepers Association for her assistance in my using the Moir Library. Lastly, to my wife for looking after me so fully thereby providing the leisure time necessary. She is also due thanks for proof reading the manuscript although often despairing of my sentence construction, which she has not amended.

In the latter part of the nineteenth century, Professor A. J. Cook, an American Entomologist, in an apology before criticising three bee books, one of them by my subject wrote:

> When any man writes a book with true, unselfish, disinterested purpose. Aiming to benefit, refine and exalt those for whom he writes, well may we call such a man a philanthropist. He is adding another sheaf to the great garner of truth, honest work, and therefore blesses the World.

Somewhat 'over the top'. However, here is my sheaf, but the very first query of sixteen years ago remains unresolved.

Chapter I

Scene setting.

Evidence has been found that humans were keeping bees domestically using manufactured containers as early as 2400 BC and the Natural History writings of the Classical authors indicate that by 1000 years ago, most materials had been used to accommodate the honey-bee. However it is likely that early man was more a honey-hunter than a keeper of bees. As with most things that have been in existence for a very long time development of the craft of bee-keeping was slow. Although the classical Greek and Roman writers, Virgil, Varro, Columella, Aristotle, Pliny, wrote about bees and bee-keeping, their manuscripts were available to few. The rate of change accelerated with the introduction of the printing press in the middle of the fifteenth century allowing relatively inexpensive access to the written word. The manuscripts of Virgil et al were amongst the first to be printed.

The material used for the construction of hives, reflected what was readily available in each locality, and the associated system of managing the bees handed on by word of mouth and local teaching. In Britain, pre 15th century the material easily available was straw, heather, reed etc., which resulted in the almost universally used skep. It was robust and readily made by the rural population using free material from the heath, hedge, or harvest field. All it needed was means of shedding the rain and a board or flat stone to put it on.

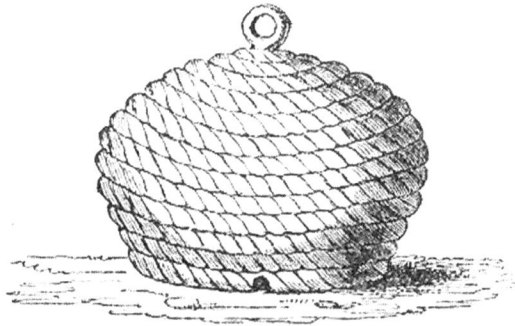

Early skep

The system of management was crude and inhumane. It relied upon the bees regularly swarming, the limited space available to them in the skep encouraging this behaviour. Hopefully, the swarms were caught, re-housed in an empty skep, placed in the position of the hive from whence it issued, moving the displaced skep to a new location in the apiary, where hopefully it would breed a new queen and prosper. At the end of the season, the hives were reduced to the original number, by taking the lightest and heaviest of the hives and placing them over a brimstone pit, killing the bees and allowing access to the honey.

The early English writers on bees did not initially address the real issues with this method of managing honey-bees. Edward Southern in the first English book completely on honey-bees, published in 1593, was more concerned with methods to encourage the bees to accept the skep they were put in, his recommendation was to:

> ....put barley, pease or malt (best) into the skep and let a hogge pig or sow eat it, turning it around as he eats so that the froth from his mouth remains in the hive, wipe over with a cloth.

I suppose that we must be thankful for the cloth! His skep was "no larger than a half bushel". Sixteen years later the Rev. Charles Butler in his much-acclaimed book *The Feminine Monarchy* described his skep in detail, which was a full bushel. The British National hive of today is twenty per cent larger. In practice, the working class bee-keeper was likely to keep bees in anything that he could find that in any way 'served the purpose'.

The publication in 1655 of *The Reformed Commonwealth of Bees* By Samuel Hartlib, provided evidence of a significant move forward in hive design and bee management in Britain. 'Reformed' in the title indicates that it was sponsored by Oliver Cromwell's government, but Hartlib was not a bee-keeper or in this case an author because the book consists essentially of correspondence he received from men who did keep bees, and develop hives. It contained details of the wooden octagonal hive, from the Rev. William Mew, the upright barrel shaped hive, by Dr. Brown, a divine, and the long recumbent hive, believed to be German. All obtained the honey without the destruction of the bees, but it was only Mew's hive that progressed in Britain, developed amongst others by Christopher Wren. Essentially they were all simple boxes and with a system of management that added extra boxes as the bees expanded and required additional space.

There followed a succession of books written around each author's hive and 'new' system of management, generally each one marginally developing the hive and improving the way in which the honey-bee was housed and dealt with. There were all shapes and sizes, with or without bars, frames, windows, but progress was very slow. The study of the insect, to obtain a better understanding of its behaviour, carried on in parallel to development of the hive and the furniture therein, and was aided by the design and manufacture of observation hives.

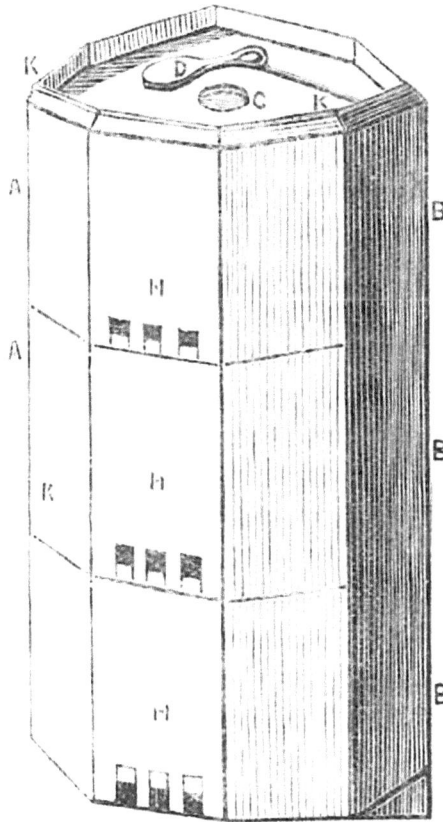

Wren's hive

Although researchers were beginning to converge on a solution, the problem still remained, bee-keepers had no adequate means of managing their bees by observation of the behaviour inside a full size working hive. In the event the solution was found by accident rather than disciplined research, and from an unlikely individual. L. L. Langstroth was an American cleric, teacher, amateur beekeeper and manic depressive, who in 1851 was keeping his bees in Bevan style boxes – simple boxes with bars, expanded vertically when required. He became frustrated that the cover board (a simple piece of wood covering the top of the hive), was always stuck to the top of the bars when he tried to remove it to get at the combs inside. To attempt a solution to this problem he modified the box to position the bars 3/8" below the top of the box, creating a space between the top of the bars and the cover board and found that the bees respected the gap, did not fill it with comb or stick it up with propolis.

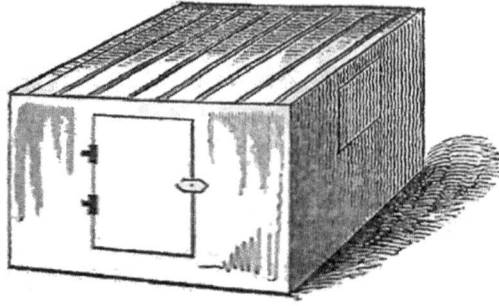

Bevan's box

Following discussions with a visitor, Langstroth realised that if it worked at the top of the hive it would also work down the sides of the comb. If he added side pieces to the top bar in such a position that they were 3/8" distant from the sides of the hive it would avoid having to cut the comb away from the hive sides. Fortunately he kept a Journal and on October 30th, 1851, he drew the sketch shown and wrote:

> If the slats are made so that *a* and *b* are about three eighths of an inch from the sides of the hive the whole comb may be taken out without at all disturbing it by cutting.

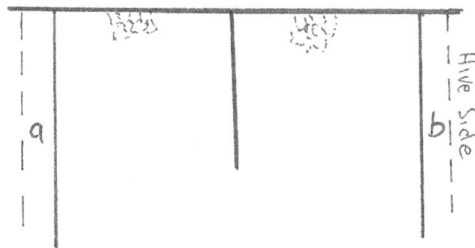

Sketch from *Life of Langstroth* by F. Naile

Langstroth had devised the first workable bee hive utilising 'bee-space' – the three eighths of an inch that the bees instinctively preserve. It is perpetually stated that he was the inventor or discoverer of 'bee-space', which is of course nonsense, others had taken note of it earlier but arguably had not managed to include it in a practical hive design, and as most centenarian bee-keeping comedians point out, the bees had always known about it, but had not been taught to write or use a saw and hammer.

On October 5th, 1852, he was granted a patent on his hive and had already one hundred of his hives made to sell or use in his own apiary. However, there followed a further short bout of his 'head demons' as he termed them, and it was only with the encouragement of friend Samuel Wagner that he set about writing *Langstroth on the Hive and the Honey-bee a Bee-keeping Manual,* which was published initially in 1853. Although one of the classic bee books, some of it is very difficult to follow with whole passages incomprehensible, the lack of illustrations not helping and when the Root company reprinted it in 1914 some of the text was changed and several illustrations from the second edition included. However, Langstroth's book, containing his hive and the complete system of management that went with it, was the basis for beekeeping from then to the present day.

The book contained two plates tipped-in before the title page; the first was a general view of the hive and is shown below, the second contained illustrations of the queen, worker and drone bees. The title page is shown and carries the classic engraving of the queen bee and her attendants, which would, in time be subject to controversy. Both these pictures are taken from my much-prized first edition of Langstroth's book.

Langstroth frontispiece

Title page of first edition

The book started with the following advertisement:

Each comb in this hive is attached to a separate moveable frame, and in less than five minutes they may all be taken out, without cutting or injuring them, or at all enraging the bees. Weak stocks may be quickly strengthened by helping them to honey and maturing brood from stronger ones; queenless colonies may be rescued from certain ruin by supplying them with the means of obtaining another queen; and the ravages of the moth effectually prevented. As at any time the hive may be readily examined and all the worms &c., removed from the combs. New colonies may be formed in less time than is usually required to hive a natural swarm; or the hive may be used as a non-swarmer, or managed on the common swarming plan. The surplus honey from the interior of the hive on the frames or in upper boxes or glasses, in the most convenient, beautiful and saleable forms. Colonies may be safely transformed from any other hive to this, at any season of the year, from April to October, as the brood, combs, honey and all the contents of the hive are transferred with them, and securely fastened in the frames. That the combs can always be removed from this hive with ease and safety, and that the new system, by giving the perfect control over all the combs, effects a complete revolution in practical bee-keeping, the subscriber prefers to prove rather than assert. Practical Apiarians and all who wish to purchase rights and hives, are invited to visit his Apiary, where combs, honey and bees will be taken from the hives; colonies which may be brought to him for that purpose, transferred from any old hive; queens and the whole process of rearing them constantly exhibited; new colonies formed, and all processes connected with the practical management of an Apiary fully illustrated and explained.

Those who have any considerable number of bees, will find it to their interest to have at least one moveable comb-hive in their Apiary, from which they may, in a few minutes supply any colony which has lost its queen, with the means of rearing another.

The hive and right will be furnished on the following terms. For an individual or farm right, five dollars. This

will entitle the purchaser to use and construct for his
own use on his own premises, as many hives as he chooses.
The hives are manufactured by machinery, and can probably
be delivered, freight included, at any Railroad Station in
New England or New York, cheaper than they could be made
in small quantities on the spot. On receipt of a hive the
purchaser can decide for himself, whether he prefers to
make them, or to order from the Patentee. For one dollar,
postage paid, the book will be sent free by mail. On
receipt of ten dollars, a beautiful hive showing all the
combs, (with glass on four sides,) will be sent with right
freight paid to any national railroad station in New England
or New York: a right and hive which will accommodate two
colonies, with glass on each side, for twelve dollars; for
seven dollars, a right and a well made hive that anyone can
construct who can handle the simplest tools. In all cases
where the hives are sent out of New England or New York, as
the freight will not be prepaid, a dollar will be deducted
from the above prices.

Langstroth clearly thought this was the path to great riches, but he failed to take account of the fact that beekeepers are born with an inbuilt mechanism that avoids parting with money if at all possible! They immediately started putting the frames into all different shapes and sizes of boxes, believing that this circumvented the patent as written. We will never know if the situation would have been improved if he had engaged a professional to produce the patent application, because effectively he needed to patent an idea. When the Rev. Homer King challenged the patent it was on the basis that Langstroth was not the first with the idea, and much later Langstroth accepted that individuals in Europe had been first but he was unaware of it at the time. Whilst he was never to reap the financial rewards he considered his by right, the honour has always been his.

The second edition of his book appeared in 1857, considerably revised with many illustrations and a new title – *A Practical Treatise on the hive and the Honey-bee,* is a much better attempt. The third edition, and last to be revised by the author himself, appeared two years later. One of the plates from the later editions is the iconic three-dimensional engraving of the hive shown.

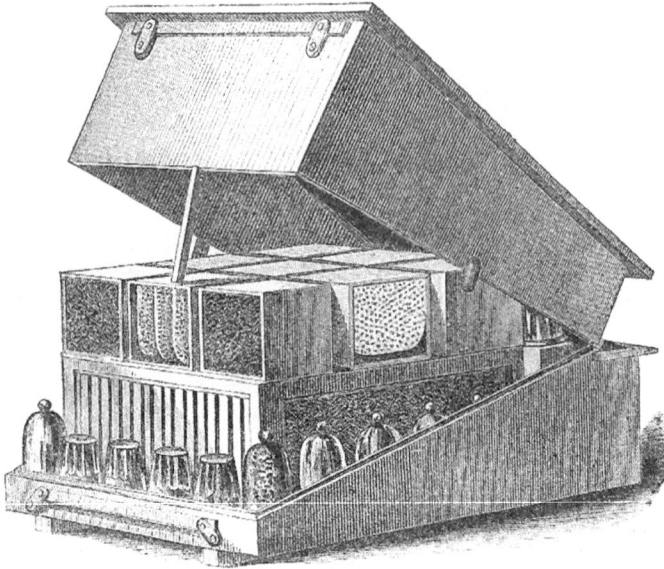

Langstroth's hive

So 1853 marked the beginning of a new era of beekeeping. The initial problem was how to convey information regarding the new principles and hives to the beekeepers in Britain. The follow-up challenge would be to persuade them that the advantages outweighed the cost of changing over; they were wedded to their various forms of ornate boxes, and beloved skeps. Additionally the cottager (working class) beekeeper had no spare money and most were illiterate.

Although some newspapers carried occasional items on bees and beekeeping, the Times having a regular, albeit inept, 'beemaster' – Dr John Cumming, it was not until the launch of *The Cottage Gardener* in late 1848, that the opportunity arose for regular information. In the very first issue the publishers stated that they wished to have a beekeeping section and asked for someone to write/edit it. J. H. Payne who had written a small book *The Apiarians Guide* in 1833, (*Beekeepers Guide* from third edition 1842), took on that role.

The first correspondent to give details of a hive that he had devised using 'bee space' was William Bernhard Tegetmeier, sometimes described as Charles Darwin's assistant, due to the large amount of correspondence between the two on a wide range of natural history topics. Tegetmeier was a naturalist, journalist, pigeon fancier, lecturer, expert on poultry and bees etc. He wrote for *The Field* from 1864 to 1907, founded the Royal Entomological Society and was secretary of the Apiarian Society of London. He contributed articles to the Cottage Gardener where he was the first to describe a bar-frame hive utilising 'bee-space', and in 1860 wrote *Bees, Hives and Honey,* a twenty page booklet, promoting the principle of the new style of keeping honey-bees. It re-appeared five years later, revised and enlarged. Initially his boxes were 11 ¾" square (internal measurement), 7" deep for the brood and 5" for the honey box (super). Later he changed this to 13 ½" square by 11" deep containing eight frames. It is interesting to speculate what might have been if Tegetmeier had made bees rather than poultry his number one interest; the B. S. standard frame would probably have been different and it is unlikely that this book would exist! His hive did not endure, and in bee-keeping he is better known for his work on cell structure.

In 1861, *The Cottage Gardener* became *The Journal of Horticulture*, and the bee section had gained another significant contributor, Thomas White Woodbury, of Exeter, Devon, who wrote under the pseudonym 'A Devonshire Beekeeper', and followed Payne as editor of the bee-keeping column until 1871. He described his hive in great detail, in the magazine. It started life as just a bar hive, but was later converted to take a frame. Again it was a simple box arrangement, this time 14 ½" square (internal) and 9" deep. Woodbury initially adapted his bar hive by devising the compound frame, which proved unsatisfactory and he soon abandoned it in favour of the simple frame. This was the hive that made the big impact in Britain and it is a shame that Woodbury did not put his undoubted journalistic talents to a book which would have had greater durability than his magazine contributions, although his description of his hive in *The Journal of Horticulture* was more than adequate.

Tegetmeir and Woodbury not only had their correspondence with Darwin in common, but they were continually at loggerheads with Cumming, many column inches occupied in several different publications on all aspects of bees and bee-keeping.

Thomas White Woodbury

Woodbury hive

Another regular contributor to the columns of *The Journal of Horticulture* was Henry Taylor who had written *The Bee-keeper's Manual* first published in 1838. The fifth edition appeared in 1855, and does not contain any bar-frame hives, and although the sixth edition in 1860, contained "with a description of the most approved hives" in its title, he fails to describe a bar-frame hive. He does, however, mention the frame and the principle without using the term 'bee-space' or as was usual for him, giving any dimensions. He is very negative about it, and appears not to have had the energy to investigate it. It was only with the seventh edition of 1880 revised and enlarged by Alfred Watts following Taylor's death, that there was a section on bar-frame hives.

Progress towards making the bar-frame hive the norm in Britain was painfully slow, the craft lacked direction, organisation and above all a leader.

The impetus came in the unlikely person of Charles Nash Abbott, son of a wealthy Hanwell builder, who had first kept bees in a skep as a teenager but they died because of lack of attention. Charles started his working life as an apprentice butcher, eventually opening his own shop. When his father died in 1863, Charles inherited one fifth of the estate, relinquished his business and went to teach in the Central London District School in Hanwell. Whilst at the school he came into contact with a Dr. Coster, the school Medical Officer and an expert on bees and bee-keeping who was already using the bar-frame hive. Abbott's interest in bee-keeping was rekindled, and with Coster he started to experiment with Langstroth, Woodbury and Quinby hives, as well as developing and making his own version. He was working in the school from

Charles Nash Abbott and George Hobbs, carpenter

C. N. Abbott later in life

6.30 in the morning to 6.30 at night, seven days a week, making hives at night and keeping forty stocks of bees. When Coster died in 1870 Abbott bought all his beekeeping equipment and it occurred to him that with the move to new hives there was an opportunity to make money from bee-keeping equipment.

Abbott started writing regular articles for *The English Mechanic and World of Science* and *The Journal of Horticulture*, advocating the Frame system of bee-keeping. He also wrote a series for his local paper the Middlesex County Times, *Hints on Bee-keeping*. He followed this in the same organ, with an attack on the members of the upper class because they took no interest in the craft of bee-keeping and not one of them was prepared to become leader or involved. The wording of the letters indicates that he had approached certain individuals without success.

Because of the apathy he experienced, in 1872 he left the school and set about establishing a National bee paper. He spent £5 advertising for those interested in bee-keeping which produced just three replies. He then wrote to 40 individuals who had written on the subject and this produced 2000 names, 60 of whom promised to pay 10/6 the annual subscription. Eventually 6000 circulars were distributed.

On 1st May 1873 Abbott launched the *British Bee Journal*. The very first issue of 16 pages plus advertisements, was of 3000 copies, 2500 being 'sown broadcast'. Of the second issue 1500 copies were sent to clergymen, and 500 to agriculturists, this produced a subscription list of about 200. The magazine was to be monthly, cost 6d, post free, or a subscription of 10/6 for the whole year, which included receiving instant replies to queries in addition to the magazine. Fortuitously it would also provide useful copy. He was to modify this structure later.

The first issue of the first magazine dedicated to bee-keeping in Britain, had a separate single sheet included, but not bound in, containing a contribution from 'H. W. T.' with a proposition for a Bee-keepers Guild. Abbott headed it thus:

> We received the following after the principle matter in our columns had been set up; yet, believing its object worthy of discussion, we supplement our pages to bring it under notice as soon as possible. A Central Guild, with branch Conventions, Clubs, or Lodges, would doubtless lead to great improvement in bee-culture here. Why are we so satisfied with a third or fourth rate position as Apiculturists?

British Bee Journal, No.1, Vol. I, page 1   Supplement to No. 1, Vol. I of *British Bee Journal*

The reverse of the article contained a full-page advertisement, which no doubt went some way to covering the cost of the extra sheet; the charge for a full page advertisement was three pounds at the time.

So the seed was sown that would eventually grow into the British Bee-keepers Association. What 'H. W. T.' made of the eventual outcome he kept to himself having said that he was unable to take part in it.

Half of the first issue of *The British Bee Journal* was, as would be expected, written by Abbott, but as each edition was published his portion reduced. Copy from others soon became available and the number of pages were increased. In the February 1874 edition, Abbott wrote an article describing his new 'Frame-bar Hive' and the following month there was an advertisement for the same hive manufactured by George Hobbs, a carpenter and joiner, who lived in Hanwell. Basic price 35/-, double skinned 50/-. By November of the same year C. N. Abbott had a full-page advertisement headed 'Bees, Hives, Bee Furniture' selling items he had previously described in detail in the magazine. Abbott had achieved his original aim.

ABBOTT'S FRAME-BAR HIVE.

GEORGE HOBBS, Carpenter and Joiner, of Wingrove Terrace, High Street, Hanwell, W., having, by permission of the Editor, undertaken the manufacture of Hives on the principles laid down in the columns of the *British Bee Journal*, and in accordance with the directions there given, will be glad to receive orders for the same. This hive has been acknowledged by all the bee keepers who have seen it, to be the most perfect, yet simple hive extant, possessing as it does all the ascertained advantages of other hives, without any of their defects.

The hive complete, including stock boxes, stand, moveable floor board, honey board, two supers, and super cover, roof, as illustrated, manufactured of mill sawn material, as described, unpainted, 35s.

The same, with moveable outer casing to the body box, with dead air space on all sides, all outside work nicely planed, and wrought, 50s.

All the parts of the hive, cut to fit, and labelled, may be put together by any amateur carpenter, including the frames and moveable side pieces, ready fitted; in fact complete patterns for amateur hive makers use, 30s.

Post Office Orders should be made payable at the Hanwell Post Office, or stamps be deposited with the Editor of the *British Bee Journal*.

First Abbott advertisement

In the November 1876, *British Bee Journal* the following notice was posted:

Mr C. N. Abbott respectfully begs to inform his friends and patrons that he has relinquished the Hive and Bee business in favour of his Sons, and that premises have been taken at Southall, where they will carry on the trade of Hive manufacturers and Bee dealers; and, the establishment being sufficiently commodious, they will open a *School of Apiculture,* where at any time, on payment of a small fee, the whole art and mystery of Bee-keeping will be communicated. The address will be, after January 1st, Abbott Bros., Fairlawn, Southall Middlesex; but in the meantime they will continue at Hanwell, W. London.

This was Charles attempt to deal with the ever-increasing criticism coming from certain individuals in the B. B. K. A. but it was never likely to quieten the noise. His business had also outgrown the space available at Hanwell. His previous offer to publish any other manufacturers descriptions of their bee equipment in his magazine, submitted with engraved blocks, had limited success. There was a faction that wanted the Association to have its own magazine "untramelled by commercial interests", although they were not prepared to fund a separate organ and Abbott was not about to give away something that he had invested money and energy in. The tensions ultimately led to Abbott resigning from the B. B. K. A. in 1877. He felt betrayed in particular by two individuals, whom he considered had been brought to prominence in the Association, by him and his magazine. But it was a complicated three-way split. Abbott considered the B. B. K. A. had become moribund and proposed starting a new Association. By March only twenty members were wishing the central organisation to continue. Of those twenty, Rev. H. R. Peel, importantly a member of a wealthy aristocratic family, was elected Honorary Secretary and duly met Abbott with a view to reconciliation. Abbott rejoined the Association one month later, the B. B. K. A. agreeing to drop the idea of starting their own magazine. Rev. Peel was not a beekeeper, but that did not stop Cowan featuring him in the series on 'Eminent Beekeepers', in 1889.

Rev, H. R. Peel

Much later Cowan revealed that "I was wishing to resign, as I found it difficult to manage the troublesome members but was induced by one or two to continue". Ideal qualities for a Chairman.

The internal rivalries were never going to disappear and as always is the case, it inhibits any organisation from functioning positively. Abbott was very energetic in his business and very forthright in his magazine, quarrelling on the size chosen for the standard frame, (Cowan); credit for authorship of *Modern Beekeeping*, (Cheshire); frame spacing, (Cowan); treatment of foul brood, (Cowan and Cheshire), etc. At the time Cheshire was editor of the bee column in the *Journal of Horticulture*. In 1882, Abbott once again, was using his magazine to verbally abuse the B. B. K. A. and their lack of effort. In the October quarterly meeting when propositions by another member regarding the operation of shows, which Abbott considered very relevant, were 'kicked into touch', he wrote "bee-keeping has become too large a subject in this country to be longer a matter of child's play".

A letter written by Abbott following the quarterly, was considered by the

Association committee at its meeting in November. Cowan answered the points raised by Abbott in a letter the same day; obviously prepared earlier. He reminds Abbott that he was lately a member of the committee but voluntarily resigned. He claimed that the committee meet the members twice a year and their representatives at four quarterly meetings, and therefore respect and represent the wishes of the members. This was a common response from Cowan, but he never appeared to appreciate that the second part of sentence does not necessarily follow from the first.

But it mattered little, Peel had already approached Abbott and offered to buy the *British Bee Journal* for £2,000, a very large sum for a ten year old magazine with a very low circulation, but Cowan later remarked that it was a small amount of money for Peel to part with.

Abbott was to edit one more edition of the magazine he single-handedly launched. The editorial was naturally tinged with sadness and a certain regret, but it did not stop him from responding vigorously to Cowan's letter, it would be his last opportunity. In brief these were his points.

1. Points out that the committee continue to be lax in the rules for showing – equipment that is never sold for he price given at the show – allowing the same equipment to win year after year.

2. The committee refuse to lay down the means of spacing frames or the distance for that. The Abbott frame spacing idea is the best about, but one member of the committee, said he would cut the ends off if that became the standard. Abbott's argument is that it helps the cottager.

3. Having pointed out errors in *Modern Beekeeping* they were ignored by the committee. Basically they cannot admit to being wrong.

4. The erroneous judging is a clear indication that judges need to be directed – only improvements in equipment to be judged – guidelines to be drawn up. This is the responsibility of the B. B. K. A.

5. At meetings where the members are present the committee normally outnumber them, making them irrelevant.

6. Committee chosen effectively from those that subscribe 20/- per annum – not likely to produce the best possible representation of the members.

He concluded by saying the remarks were in the interest of beekeeping and realised he ran some risk; but Abbott was no fool.

At this time Abbott Brothers employed twenty-five men, two years later that had doubled, the firm evolving to eventually making 'Abbess' office furniture. In 1886 the three brothers were reduced to two when James left to set up his own bee business in Ireland.

Following a letter to *The Standard*, a testimonial fund was set up for Abbott in recognition of his services to beekeeping. F. Cheshire did not contribute; T. W. Cowan did, and in a break with tradition Abbott's closest associate, R. Symmington, donated twice the amount of the President, Burdett-Coutts. £37 was raised and at the April 1883, conversazione, Abbott was presented with a 'handsome black marble' dining-room clock and an illuminated address on vellum, including the names of those who contributed. Neither F. Cheshire nor T. W. Cowan were present, the latter citing the 'serious illness' of Mrs. Cowan, not the last time that this would be used to excuse attendance.

Charles Nash Abbott with the assistance of a few friends and close relatives, had worked to create the framework for bee-keeping to exist and thrive in Britain, and for his family to create wealth. All those that followed, including my subject, owe their fame, whatever it was, to Abbott. Charles Nash Abbott died on 2nd March 1894. Three weeks earlier when his death was known to be imminent, somewhat reluctantly, he was made an honorary member of the B. B. K. A.

Chapter II

Genealogy

Thomas William Cowan was born in Leningrad, Russia on the 2nd January 1840[1] to a Scottish father John, and an English mother Mary Anne (Eales). John and Mary Anne had married on 10th November 1832 in the British Chaplaincy, St Petersburg. Thomas was the first surviving child of the marriage, Demitrius followed two years later when Mary Anne was 42 years old, completing this Cowan family. Mary Anne is believed to have been John's third wife; he had one child Catharine, from a previous marriage.

It was claimed that Thomas was "a member of the famous family of paper manufacturers of Edinburgh", but much research by the authority on this family, has been unable to verify this. Further investigation shows the probability that his father, John, was born on 6th January 1795 to Thomas, a lorimer, (maker of items for dressing horses), and Catharine (Tosh), his wife, when they resided in Causewayside, Edinburgh. Of course it is possible that this Thomas was a very distant relative of the famous paper–making family. The Cowans are a sept of the Colquhoun Clan who originated in Dumbartonshire close to Loch Lomond. Mary Anne originally came from Clerkenwell. The census information is confused about her birth year, but I believe this is most likely to be 1799. She was also known as Ann and Annie causing further uncertainty.

In St. Petersburgh, John worked as a Civil Engineer, believed to be for one of the Scottish Engineering companies operating there at the time. He was there from 1818 and reportedly held the rank of General. It is not known exactly what he did, but this was a time when British engineering was being exported throughout the World, and the likelihood is that he was involved with manufacture. He was not alone; there were many Scottish engineers employed by famous name Scottish Companies in Leningrad at the time.

The Cowan family returned to England at the outbreak of the Crimean war in 1854. Two years later, following the death of William Joyce, John took over his engineering business, Kent Iron Works of Greenwich. In the 1861 census the Cowan family are shown living at 20, Surrey Terrace, Deptford, Kent, with a Cook and Housemaid. John is listed as a 'fund holder', Thomas a Civil Engineer and Demitrius employed as a clerk to a Ship and Insurance Broker.

---

1 This is the date given in the International Genealogical Index (IGI), which also gives his date of christening as 2nd April 1840. Because Russia did not change to the Gregorian calendar until 1918, these are dates from the Julian calendar. His headstone gives the date of birth as 14th January 1840 which is the equivalent date in the Gregorian calendar. 1582, Pope Gregory XIII took ten days out of the Julian calendar at the first change, each additional 138 years adds another day. 1840 – 1582 = 258, approximately two additional days – twelve.

With John listing himself as a fund holder the implication is that Kent Iron Works had already been transferred to Thomas who would have finished his apprenticeship by then.

Following an introduction effected at the Country residence of Joshua Field, on the 11th of May 1864, Thomas married Fanny Matilda Michell in Horsham, Sussex. Joshua Field was a founder member, later President of the Institution of Civil Engineers, and partner in the firm of Maudsley, Sons & Field of Lambeth, a company that John Cowan had done business with. It is also likely that Thomas and Fanny had encountered one another at the 1862 Great Exhibition at Alexander Palace, to which the Michell family held season tickets for parents and their two children, and at which Kent Iron Works, exhibited. Their first home together was Rhyde House, Beckenham, Kent. Fanny was Thomas's senior by just a few days.

Thomas was five years younger than the usual age for aspiring gentlemen to marry because it was considered that members of this class would have money in the bank before proposing, it was the age of investment income, savings, and best of all unearned income. It is doubtful if Thomas had much in any of these categories, so he must have done a very good job in convincing Henry Michell that his prospects were much brighter than his bank balance. As the years pass there does appear to have been genuine affection between Thomas and Fanny and they did much together, unusual in the male dominated society that characterised the Victorian period.

Fanny's father, Henry Michell was a very successful, wealthy Horsham businessman, and as a consequence, an important local dignitary. He ran a brewery, as did his father before him, owned and operated many public houses, several brickyards, and for a limited period a coal business; his investments were many including water and railway companies. Fanny's mother Sarai (Ellis) was from a local land owning, farming family; her sister married one of Henry's brothers. The diaries of Henry Michell fortunately survive and his first entry mentioning his future son-in-law read:

> We are all looking forward to the marriage early in the new year of our dear Fanny to T W Cowan to whom she has been engaged for some time and perhaps this is the proper place to state that he was born at St Petersburg his parents were in the office under the government there and had been Mr Cowan ***** for between 30 & 50 years they were compelled to return to England at the breaking out of the Crimean war and as they were often staying with Fields at Tonbridge we became acquainted with them there.

This shows that Cowan had rented the house in Beckenham and in 1970 moved into the Horsham property that appears to have been rented by Henry Michell for the Cowan family.

In the 1871 census Thomas and Fanny are registered at 32 South Street, Horsham, (Hawthorne House), with Alexander, a cook, housemaid and nursemaid. Thomas is listed as a "Civil Engineer, Ph. D., M. A. etc".

Thomas and Fanny had four further surviving children: -

Helena Maria, 1871

Edith Constance, 1872

Herbert Francis, 1874

Percy John, 1876

All four of these births were registered in Horsham.

Henry Michell senior died on the 25th October 1874. His will of 17th October 1872, and the two subsequent codicils of 28th February 1873 and 23rd October 1874, were proved on the 4th February 1875. In it he established an investment trust with £25,000 from which Fanny was to receive the income, the capital:

.... shall be free from the control and engagements of her present or any future husband and her receipts alone for the said monie when and as the same shall from time to time come due.

Fanny would get the return on the investment, £750 - £1,000 per year, and Thomas could not touch any of it. If Fanny died before Thomas he was to have the income from £5,000 of the trust fund, the income from the remaining £20,000 to go to each of Fanny's children as and when they became 21 years old. The will covered every possibility, including death of Fanny in childbirth, a real sign of the period. Henry had attempted to leave a legacy to all descendents of his daughter and although he probably thought it unlikely, the possibility of the line ending was also covered, the capital then returning to his main trust.

In the first codicil Thomas is bequeathed £300 per year from the time of Henry's death to the end of the existing lease (from Sir Percy Shelly) on the brewery in 1886. This was the only substantial addition to the main will, the other items being small donations to relatives and employees. This appears to imply that Henry expected his invested money to return 6% and not the 3% - 4% I have worked to above, but it gives no indication of the reason for the change of mind that led to the inclusion of Thomas as a benefactor.

As with all such wills it is impossible to ascertain what happened with the passage of time. Sarai Michell died in June 1878.

By the time of the 1881 census, the Cowan family are to be found residing at Comptons Lea, Union Lane, Horsham, a mansion in forty acres that Thomas had built and which the family moved into in 1878. Could a Michell company have supplied the bricks? Alexander is away attending the Grammar School in Berkhampstead, Hertfordshire, Headmaster Rev E. Bartrum, (one of the early committee members of the B. B. K. A.). The remainder of the family is at home. Additionally there was a visitor, Governess, Cook and two Domestic servants. With a combined income of between £1,000 and £1,600 they were 'comfortable'. Thomas is listed as a farmer of forty acres, the implication being that he was no longer operating the Kent Iron Works and probably had not been for more than a decade. The farm should have produced a small additional income. Following a chequered career as a private residence, hospital, and flats, Comptons Lea was demolished in 2002 and a block of flats was built on the site.

Architect drawing for Comptons Lea

Photograph of the other side of Comptons Lea, just prior to demolition in 2002.

Although he was educated and trained as an Electrical Engineer, Alexander emigrated to America in 1895 and set up as a fruit farmer at Two Pine, Placer, California, adjacent to Loomis, a major centre at the time for transporting and trading fruit. In America he met, and in 1900, married Frances, eight years his junior, who had emigrated from Ireland in 1891. The happy event was reported by Broughton Carr in the *B. B. J.* on February 8th, 1900:

> Marriage Of Mr A. H. Cowan.
> We have just received from our senior editor the pleasing news that his eldest son, Mr A. H. Cowan – who has a fruit farm in California, growing oranges and peaches on a large scale – was married on the 9th ult. At All Saints' Church, Loomis, Placer, California to Mary Owen, eldest daughter of Major G. H. Turner, and grand-daughter of J. J. Turner, Esq. of Penthreheylin Hall, Montgomeryshire. For the benefit of Mr Cowan's Australian friends we may also add that the lady is a grand-daughter on the maternal side of the Hon. S. Tomkinson, M. L. C. of Adelaide South Australia.
>
> The many readers of this journal who, along with ourselves have known Mr. Alec H. Cowan personally, only to esteem him very highly, will assuredly join us in congratulating the young people on what bids fair to be a very happy future for them. If the well known augury counts for anything, this should be certain for we learn that the wedding day was one of brilliant sunshine, coming between two of the wet and gloomy days so rarely seen in that region of almost perpetual summer.

Although he was named Alexander, the family appear to have used Alec. Also the information in this notice in respect of his bride does not concur with that given earlier, the source for which was American immigration records and censuses, the confusion being resolved on her headstone where she is named Frances Mary Owen.

Alexander and Frances had no children. Early in the 20th century they returned to England, and lived with his parents at Upcott, Bishops Hull, near Taunton, Somerset, until Thomas and Fanny left for Clevedon in 1918, when son and daughter in law moved to 'Loomis', Milverton, Somerset. Until his health failed, Alexander was proprietor of the Taunton Vulcanising Works, Castle Green and Bridge Street. There is some evidence that he gave up the business in 1924. During WWI he was the army recruiting officer for Somerset, holding the rank of Captain, and for which he was awarded the O. B. E.

He was church organist at both Bishops Hull and Milverton. He died on 30th April 1931; Frances died on 2nd March 1947.

Helena appears with the rest of the family in the 1881 census, but is not at home with them in the 1891 census when she would have been 19 years old. At the time of the census she is listed as a visitor at Criagallian (Craigallion), Strathblane, Stirlingshire, Scotland, the home of Allan G. Barns Graham, described as landed proprietor. The Grahams had descended from forefathers that had been given their land by William the Lionheart. Prior to returning to Craigallion, Allan had been an East India merchant in Manchester. It is difficult to determine what possible connection the Cowans would have had with a Scottish landowner that would have resulted in Helena being invited there, but there were several young men in the family of a similar age. It does appear that the only habitable residence at the time would have been Westburn, a three storey foreboding property, that a few years later was demolished when the Cambuslang Golf course was laid out in its grounds.

Herbert is listed with the family in both 1881 and 1891 censuses. In the latter he is given as an Electrical student, then aged about seventeen.

Helena and Herbert were two of the forty-one passengers drowned when travelling to New York on the SS Mohegan that inexplicitly smashed at full speed into the Vase Rock, one of the Cornish Manacles, just before seven in the evening of 14th October 1898. The SS Mohegan, a very luxurious ship of 4500 tons, sailed from Tilbury on 13th on what was only her second voyage, having been purchased new by The Atlantic Transport line a few months earlier. On her maiden voyage from London to New York she had leaked alarmingly and was returned to the river Tyne for a refit. On the second voyage leaking boilers had been reported immediately she sailed. Neither of these factors could have contributed to Captain Griffiths, Commodore of the Line, steering completely the wrong course once past the Eddystone lighthouse. Conspiracy theories abounded at the time because Griffiths was a shareholder and in debt, but nothing would change the fact that Thomas and Fanny had lost two of their five children, aged 27 and 24. The Cowan children were travelling to a new life in America with their parents and brother. Parents and Edith were booked to travel on a different ship and route a few days later. Of 167 passengers and crew only 60 survived, 10 were passengers, all women and children, indicating an orderly abandoning of the ship. However, the passenger list, with all passengers given as foreigners, might have been an indication of the Captains state of mind on departure, or perhaps it was normal for the time. Those names crossed through survived. Both Helena and Herbert were listed as having no profession, occupation or calling.

SUMMARY of CABIN PASSENGERS.

NATIONALITIES.	NUMBER OF SOULS.							
	Adults of 12 Years of Age and upwards.				Children between 1 and 12 Years.		Infants.	
	Married.		Single.					
	M	F	M	F	M	F	M	F
English	—	—	—	1				
Scotch								
Irish								
Foreigners	4 5	10 12	12 15	10 11	3 3	—	—	~
Total								

	4	10	12	11	3	–	No. —	
Total number of Adults							4 6	
Children between 1 and 12 : equal to Statute Adults							2½	
Total number of Statute Adults							4 8½	

I certify that the above is a correct List of the Names and Descriptions of all the Passengers who embarked at the

Signed .... Master.
.... Emigration Officer

.... Officer of Customs at London.

Mohegan passenger summary

SS Mohegan

Mohegan passenger list\

The deaths were announced in the *B. B. J.* on the 20th October, 1898, thus:

With the most profound sorrow we have to announce the sad news that Miss H. M. Cowan, the eldest daughter, and Mr. Herbert F. Cowan, the second son, of our senior Editor, were passengers on board the ill-fated Atlantic Liner Mohegan, wrecked off the Cornish coast on Friday last, and that the lives of both were lost.

It would be out of place at the present juncture for us to say any more than that Mr. Cowan will be unable to fulfil his engagement to attend the meeting and conversazione of the B.B.K.A. at Jermyn-street this evening.

A week later Broughton Carr wrote:

Obituary
In the wreck of the Mohegan, near the Lizard, Cornwall, Helena Mary, aged twenty-seven, eldest daughter, and Herbert Francis, aged twenty-four, second son of Thomas William and Fanny M. Cowan, of Hampstead. " Not lost, but gone before."

The above announcement, copied from the ordinary first column of the Standard of the 19th inst., was written by our senior editor himself, and is inserted here because of some regrettable errors in the newspaper accounts regarding Mr. Cowan's connection with the disaster, which have, at least, caused misleading confusion, and given some pain to both himself and his friends. The latest instance of this—giving force to my words—occurs in the Standard of the 22nd, wherein it is stated that "the bodies of Mr. and Mrs. Cowan have been recovered." Having said this much, it devolves upon me to perform a sorrowful and somewhat delicate duty to readers as follows: On Saturday (22nd), last, I saw Mr. and Mrs. Cowan at Euston, when they left by train to join the Cunard liner Etruria, bound for New York, the travellers being en route for Loomis, California. In conversation prior to the train leaving, the question naturally arose as to what should be said to readers of the Bee Journal about the sad bereavement which had befallen the family. I was anxious myself regarding this, knowing Mr. Cowan's deeply-rooted objection to publicity for himself in print. But the latter could not be ignored, and so I thought it best to put the question by asking : " What can I say in Thursday's Bee Journal about yourself after all that has happened?" I half feared the reply would be, " Nothing," but after a moment or two he answered, " Well, say what you like, but not too much." This I will try and do, but in doing it have decided, so to speak, to leave the editorial chair and the editorial " we," and address our readers as personal friends, who, like myself, are sad enough at the pathetic break in a happy family by the disaster, but anxious to know a little more on the subject than people who are not bee-keepers. The simple facts, then, as they occurred regarding our senior editor's son and daughter on board the ill-fated steamer Mohegan are as follows : —

Mr. and Mrs. Cowan had arranged to pay a lengthened visit — accompanied by their only two daughters and second son — to America, where their eldest son owns a fruit farm, in which his father is much interested, at Loomis, California. The voyagers thus included all the members of Mr. Cowan's family except the youngest son, who is a pupil at the engineering works of the Great Northern Railway, Doncaster. Probably all would have travelled from Liverpool in the Etruria, but the two young people who were passengers on the Mohegan, being fond of the sea, chose to take the longer voyage from London and meet their parents at Chicago, Mr. Cowan himself having arranged to pay a promised visit to some American and Canadian bee-keepers on his route to that city. On Thursday, the 13th inst., he saw his children depart on board the Mohegan from the Tilbury Dock in high spirits and full of happy anticipation of their next meeting at Chicago. It was his intention to join Mrs. Cowan at Doncaster on the following Saturday and spend the intervening few days there with their youngest son referred to above, the latter being the only member of the family thus left in England. But how true is it that " Man proposes but God disposes." Saturday brought news of the loss of the Mohegan off the coast of Cornwall, and the day was spent in anxious suspense by all who had relatives or friends on board. At 9.30 p.m. I received at my home a "wire " to say that Mr. Cowan and family had gone from Paddington to the scene of the wreck, and on Monday our worst fears were confirmed. I, of course, at once wrote to Mr. Cowan, but a note from himself crossed mine, in which I learned that the worst had happened, but that he had recovered the bodies of his children, not bruised, as so many were, but placid, as if sleeping peacefully, and he was thankful. I pass over what immediately followed, except to say that they were buried on the 19th inst., at the little village church of Budock, close to the scene of the disaster, the family returning to town the same day. I saw Mr. and Mrs. Cowan at Hampstead on Thursday and found them full of gratitude for the many letters expressing sympathy with them in their bereavement. Not a few have reached this office conveying similar sentiments. One, inserted on page 428, will probably suffice as expressive of the feelings of our readers, and knowing how greatly Mr. Cowan regards the good wishes of bee-keepers, I ask them to accept, on his behalf, this acknowledgment of their kindness. In conclusion, and as Mr. and Mrs. Cowan, with their only remaining daughter, are now on the Atlantic on their way to California, where they proceed direct from New York. I hope to be pardoned for quoting a few words from a note to myself, written by Mr. Cowan on board the Etruria, after leaving Liverpool, and thus conclude.

The extract reads thus: - "I can hardly realise that so much has happened since last Saturday, and that everything has been so ordered as not to prevent our joining Alec, as soon as possible. When we think of how others are suffering who have not found their friends, or have found them mutilated, how thankful we are that ours were spared all suffering, and that we were able to recover their precious remains so soon. Now we are on the sea, but 'in the hollow of His Hand and trust in Him who doeth all things well."

The above words are more characteristic than any I could write of one for whom so many of us feel a love and esteem seldom given to a single individual in this world.

W. Broughton Carr.

The piece above was repeated in the *American Bee Journal*.

Summarising the family movements: -

13th. Thursday. Helena and Herbert leave Tilbury and drown later the same day.

15th. Saturday. Thomas was due to travel to Doncaster. News of shipwreck breaks. Mrs Cowan and Percy travel from Doncaster to London and the whole family travel to Cornwall.

17th. Monday. W. Broughton Carr received confirmation of deaths by wire.

19th. Wednesday. The two children are buried and family return to London. Notice of deaths appear in the Standard.

20th. Thursday. Death announcement in *B. B. J.*

22nd. Saturday. Mr and Mrs Cowan and Edith leave for Liverpool and on to America.

Headstone of Helena and Herbert in Budock church

When in Loomis at his son's place Cowan wrote:

PERSONAL

Dear friends, - My family and I have been quite touched by
the flood of loving sympathy extended to us by our friends.
By every mail a large number of letters have been pouring
in, not only from friends in England, but also in all
parts of the world. Nor have bee-keepers been behindhand
in expressing their sympathy – first, at the meeting of the
Council of the British Bee-keepers Association; then at the
Convesazione; and subsequently by letters from those known
to me personally, and also from many whom I have never met.

As it is quite impossible - much as I should desire it - to
answer individually all the kind and sympathetic letters we
have received from the many friends and subscribers to the
BRITISH BEE JOURNAL and RECORD, we feel obliged to tender
our thanks in this manner, and wish to assure all those who
have so lovingly remembered us, that the thought of our
being so sympathized with and prayed for has indeed been

a source of much consolation to us in our sorrow.

At first our trial did seem almost overwhelming, but very
soon we were able to recognise our Father's loving hand
in it all, by the assurance He sent us that all was well
with our dear ones. Their bodies were so quickly recovered
absolutely unharmed by the cruel rocks they had been tossed
among, and the beautiful look of holy calm upon their faces
testified that they had died as they had lived, and to their
entrance into glory, so that we could not but feel that,
although taken from us, they were present with the Lord.
So our hearts were comforted, and we felt indeed that they
were not lost but gone before.

Thanking you all on behalf of my family and myself for this
mark of your loving esteem, - I remain yours faithfully,
THOS. W. COWAN, - Loomis, California, November 16, 1898.

It would appear that the Cowans were not prepared to let the death of two of their children interrupt their arrangements. Is this the manner in which this class of Victorians dealt with the death of close family members or does it reflect the contempt of his family that some thought he had? There is little sign of true grief or emotion in the letter above.

Edith Constance was the Cowan's third surviving child born in Horsham on the 22nd November 1872. In the 1891 census she was at home and listed as an Art Student. There is little known of Edith. She, like the rest of the family does not appear in the 1901 census because she was in America with her parents and eldest brother. In the U. S. Federal census for 1900, Edith is listed with her parents in California. She lived with her parents in Taunton upon their return to England, and moved with them in 1918 to Clevedon residing with them until the end of their lives. She did not marry and is buried and death registered with the name Cowan. She died on the 26th September, 1947, and is interred in the same plot as her parents. Some writers felt that Edith was little more than a servant, and treated with contempt by her father. She appears to have been very religious and after returning from America staged an exhibition in London of 'flora and forna' from California in aid of a religious charity.

Percy John was the last child of Thomas and Fanny, born at Horsham in June 1876. In 1921 he married his second cousin, Evelyn Mary Sharpe (b. 1882), the ceremony took place in Brentford. Given their ages it is not surprising that they were childless. Percy was a Civil Engineer, and had a distinguished career as a Railway Engineer, Journalist and in the Army. Initially he worked as a Civil

Engineer on the British railways, then as a journalist with the magazine *Engineer*. At the start of WWI he joined the Army, ultimately reaching the rank of Lt. Col. when he left in 1920. His war office record gives him as Captain in 1918; he was awarded the M. B. E. for his Army service. In 1920 he became editor of *Engineer*. He achieved full membership of the Institution of Civil Engineers, who awarded him a George Stephenson medal and a Telford premium for papers that he delivered at a conference held in America. His father, despite claiming to be a Civil Engineer, did not attain membership at any level of the Institution; the real measure of an individuals qualifications and experience in that discipline. In 1936 Percy and his wife were residing at Upcott, Cranes Park, Surbiton, Surrey, where they ended their lives, Evelyn in December 1952 and Percy in November 1954. Both are interred in Bishops Cannings cemetery.

And so this particular line of the Cowan family died out and presumably the Michell family trust became £25,000 wealthier. It is interesting to note that none of the Cowan off-spring achieved the age of their parents, unusual for the time and perhaps a reflection of the comfortable manner of Thomas and Fanny's life.

Thomas's brother, Demetrius, continued living with his parents initially at 20 Surrey Terrace, Deptford, then 92, Lewisham High Road. When he married Louisa Mary Cater early in 1865, they both lived with his parents. Demetrius and Louisa had seven surviving children. John and Anne Cowan died in 1886, John at home on 1st February and Ann also at home on the 30th April. Demetrius and his family continued living at the same address whilst he continued to work initially as a clerk in a ship broking office, then in a financial institution. He survived his brother and was present at Thomas's funeral. Demetrius and Louisa spent their retirement in Falmouth, Cornwall, Louisa's county of birth, and where Thomas once rented a property, Louisa dying in early 1928 aged 87 and Demetrius in the latter part of 1933 at 92 years of age.

For the period John Cowan, his two sons and their three wives all had extraordinarily long lives, reflecting the relative comfort in which they lived; very surprising in Thomas and Fanny's cases because they reportedly suffered many bouts of 'serious illness'. Thomas William Cowan died at home on 23rd May 1926 having fallen from his library steps on the 17th.

The *B. B. J.* of June the 3rd 1926 carried his obituary, by W. Herrod-Hempsall, headed Mr. Thomas William Cowan, F. L. S., F. G. S., F. E. S., D.Sc., Ph.D. and including his latest portrait. A very modest entry by normal standards followed by a very brief account of the funeral thus:

> In the presence of a representative congregation, the funeral took place at the Parish Church on Wednesday afternoon, (May 26). The Rev A. W. Woolverton (incumbent

of Christ Church) officiated assisted by the Rev Preb. H. E. Badcock, R. D. (vicar of Clevedon), who read the lesson. The cortege was met at the western entrance by the clergy and choir boys, and Miss Philips was at the organ. The chosen hyms were "O God, our help in ages past," and Nearer my God to Thee," whilst Psalm xc was chanted, as were also the Nunc Dimittis prior to leaving the church for the graveside. During the assembling of the congregation Miss Philips played "I know that my Redeemer liveth" and "Blest are the Departed," concluding after the Nunc Dimittis with Mendolson's "O rest in the Lord".

The mourners were Capt. A. H. Cowan (son) and Mrs Cowan, Major P. J. Cowan (son) and Mrs Cowan, Rev S. L. R. Sharp and Miss M Sharp (cousins), Mr and Mrs Francis Budgett and Mrs G. M. Turner. Among the friends present were Mr. L. Snelgrove , of Weston super Mare (Chairman of the Somerset Beekeepers' Association); Mr W. West, of Bridgewater (secretary), Mr S. A. Bradbury and Mr R Beck. The unpolished coffin bore the inscription "Thomas William Cowan at rest, May 23rd 1926, aged 86 years."

A beautiful floral tribute from the family reposed on the coffin, and amongst others who sent wreaths, etc. were the following: - Mr D. Cowan and family, Miss M. Cowan, Western Division Somerset Beekeepers, British Beekeepers Association, and the Clevedon Horticultural Society.

Although other reports of the funeral differ in respect of those present, none mention representation of the B. B. K. A. although Cowan was President at his death. *The Bee World* carried a glowing obituary by Annie Betts, but later she was to change her mind about him.

Fanny Matilda Cowan died at home in Clevedon on 11th July 1928 aged eighty-eight. Spike Milligan's famous epitaph "I told them I was ill" springs to mind. She is buried with Thomas.

Wills (abridged).

Thomas William Cowan. The will was dated 27th October 1896, when residing at 31 Belsize Park Gardens, Hampstead, and drawn up by his usual solicitor – Edwin Ellis of Cook and Ellis. He named Fanny as sole executor, but in the event administration was granted to Alexander and Constance on 26th

August 1926 He left all household goods and effects, cash in hand, money in current and deposit accounts to Fanny. The remainder was left in trust to his children, Fanny being the sole Trustee, with freedom to do as and when she decided with its contents.

The Gross value of the estate was £5864 11s 1d, and the net value of Personal estate £1980 4s 10d.

It is interesting to note that he saw no need to modify his will in the last thirty years of his life, made no bequest to any organisation or individual outside the family, appearing to contradict the image he cultivated of being a very generous individual. Specifically there was no bequest to the B. B. K. A. who eventually purchased his collection of books from the beneficiaries.

Fanny Matilda Cowan. A few weeks after Thomas died, Fanny put her name to a new will, necessary after her husband's death. In this will dated 4th June 1926, naming Edith and Alexander as executors and trustees, she initially gave three small legacies to differing religious organisations. She specified that Alexander should have piano, Dan Quare clock, large microscope and slides, and her sideboard "made or designed and carved by my late husband"; did she not know which? Edith was left £100 and her choice of whatever she wanted from the household goods and effects, and personal items, the remainder of such items to be divided between her two sons. The residue of her property real and personal was for Alexander. The three children each inherited one third of the income from Henry Michell's trust set up for his daughter which was transferred upon Fanny's death, and would have been worth between £300 and £500 per year to each.

The gross value of Fanny's estate was £1337 12s 0d, the net value of the personal estate being £1238 19s 8d, the will being proved on the 30th August 1929.

John Cowan. Written on 25th March 1875, over a decade before his death, this was the simplest possible will in which he left his whole estate to his wife for her use as she desired and also made her the sole executor. The will was proved and registered on the 10th March 1886, some six weeks after his death.

The Gross value of his personal estate is given as £4724 2s 4d.

Ann Cowan. The gross value of her personal estate is given as £5,285 1s 0d. Thomas, one of the executors, was bequeathed her estate at Waddon absolutely ; Catharine, John's daughter by his second wife, was left, the income from her Russian bonds to be held in trust for the rest of her life; Demetrius was bequeathed her two leasehold houses. Upon Catharine's death the bonds were to be divided equally between Thomas and Demetrius. There were a few minor bequests. I have been unable to find what the estate at Waddon (central

Croydon), comprised, and there is no evidence that Thomas or his family ever used it, or had disposed of it before his departure to America in 1896. Thomas ultimately benefited by approximately £3,000 from his parents.

Henry Michell. This is a very complex and detailed will, made more difficult to decipher by being in small closely written manuscript. There were also two codicils. The relevant dates are: -

Will. 17th October 1872.

1st codicil. 28th February 1873.

2nd codicil. 23rd October 1874.

Death 25th October 1874.

Will and both codicils proved 4th February 1875.

In understanding his bequests it has to be remembered that he had already given half of the income from the business to his son.

Will. To his wife he left their house for as long as she lived, all the house contents, Brougham, horses and accessories, and £1000 per annum.

A trust was to be set up for Fanny, by liquidising £25,000 of assets, and she was to have the income from that trust for the length of her life, "…for her separate inalienable use….free from control and engagements of her present or future husband". If she died before Thomas he was to receive the income from £5,000 of the £25,000, the income from the remaining £20,000 being divided equally between any living children. In the event, this did not happen, and when Fanny died the three surviving children at that time inherited the income from the whole £25,000. Because none of the Cowan children had off-spring, all of the £25,000 eventually reverted to the Henry Michell family trust, making much of the detail in this part of Henry's will not applicable.

First codicil. One pound per year of employ, for all those who worked for him.

£500 to friend and servant George Handford and £50 to his son, also George.

£100 to both his brother Alfred and sister Mary. Just two selected from eleven.

£100 to the local Literary Institute.

To Thomas William Cowan £300 per annum, to be taken from the profits of the brewery and malt-house, until the lease on same expires in 1886.

There is clarification on where the £25,000 for Fanny's trust should come from, stating that shares involved need not be liquidised, but simply transferred into the trust at book value. The Trustees and executors have responsibility for any losses removed from them.

Second codicil. Revoked bequests to workers and substituted 5/- per week to two widows, one of brewery manager and the other of his brother.

One would have thought that the clarification of Fanny's trust deemed necessary in the first codicil should have been pointed out by the lawyer, when drawing up the initial will. Also questions arise as to why the individual fixed bequests in the first codicil could not have been also included in the original will. The bequest to Thomas was substantial and by its nature would have been exactly the same whenever made, so again could have been included in the will. It will always be a mystery as to what happened in the intervening 16 months to bring about such substantial changes. Also, when he had gifted money to his workers on the occasion of his son's marriage, why on his deathbed did he remove them as beneficiaries?

Henry Michell became ill towards the end of 1871, not venturing out for six months, and his diary indicates that he was beginning to think of death which probably initiated the drawing up of his will. He remained unwell until his death, but he gives no indication of circumstances that led to the changes above.

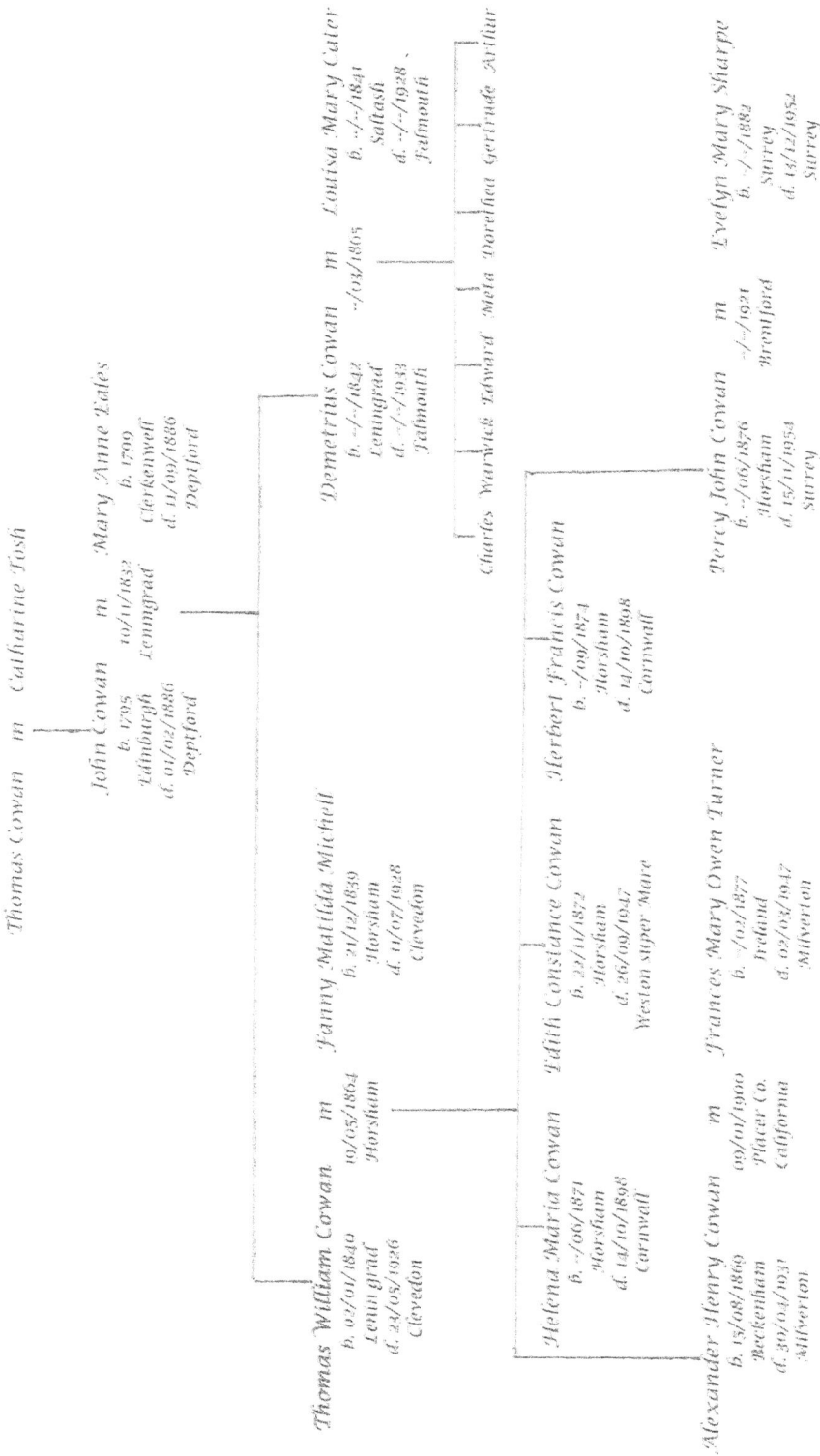

Thomas Cowan   m   Catharine Tosh

John Cowan   m   Mary Anne Eales
b. 1795   10/11/1832   b. 1799
Edinburgh   Leningrad   Clerkenwell
d. 01/02/1886   d. 11/09/1886
Deptford   Deptford

Thomas William Cowan   m   Fanny Matilda Mitchell
b. 01/01/1840   09/05/1864   b. 21/12/1839
Leningrad   Horsham   Horsham
d. 24/05/1926   d. 11/07/1928
Clevedon   Clevedon

Demetrius Cowan   m   Louisa Mary Cater
b. -/-/1842   -/05/1805   b. -/-/1841
Leningrad   Saltash
d. -/-/1913   d. -/-/1928
Falmouth   Falmouth

Charles   Warwick   Edward   Meta   Dorothea   Gertrude   Arthur

Helena Maria Cowan
b. -/06/1871
Horsham
d. 14/10/1898
Cornwall

Edith Constance Cowan
b. 22/01/1872
Horsham
d. 26/09/1947
Weston super Mare

Herbert Francis Cowan
b. -/09/1874
Horsham
d. 14/10/1898
Cornwall

Alexander Henry Cowan   m   Frances Mary Owen Turner
b. 15/08/1869   09/01/1900   b. -/05/1877
Beckenham   Placer Co.   Ireland
d. 30/04/1931   California   d. 02/02/1947
Milverton   Milverton

Percy John Cowan   m   Evelyn Mary Sharpe
b. -/06/1876   -/-/1921   b. -/-/1882
Horsham   Brentford   Surrey
d. 15/11/1954   d. 15/11/1951 (?)
Surrey   Surrey

Family tree

Chapter III

Industrial Career.

Upon their return to England from Leningrad in 1854 the Cowan family settled in Deptford. It has been reported that Thomas continued his education at Brighton College, but whilst they have complete records of students from the very start of the College, the archivist was unable to find any mention of him. The only source of this information could have been Thomas. Two years later, following the death of Engineer W. Joyce, John took over his Greenwich business, comprising a boatyard on the Thames and an engineering facility close by. The Company was Kent Iron Works.

Both Ron Brown in *Great Masters of Beekeeping* and David Charles in *Somerset Beekeepers and Beekeeping Associations A History 1875 –2005*, state that Cowan was trained as an engineer at the London School of Mines. The Government School of Mines and Science Applied to the Arts was formed out of the Sir Henry De La Bech's Museum of Economic Geology in 1851. After 1863 they split again; formal education being carried out in the now Royal School of Mines. Forty or more lectures per year cost £4 and for between thirty and forty the charge was £3. After three years of this, those who passed 1st class received a certificate as Associate of Royal School of Mines, letters Cowan never used after his name, but it is unclear if those that finished before 1863 had any formal recognition of their education.

Imperial College is the present day educational establishment descendant of the School of Mines, and its archives reveal the T. W. Cowan was registered at the Royal School of Mines, his entry stating that he was an occasional student studying Chemistry and Applied Mechanics, 1858 – 1859. He would have had to fit this in with his apprenticeship and could only, therefore, have attended in the evening. Thomas is not listed amongst the alumni of the time, a list that includes many famous names. The only other source of education in the Engineering sphere at this time was the Mechanics Institutes, but these were primarily for the working class, holding lectures in the evenings and not somewhere Thomas was likely to have frequented. Thomas's fascination with geology and his later associated museum at Comptons Lea, could well have resulted from time spent in the Museum of Economic Geology.

Thomas was apprenticed to the company owned by the Rennie brothers, George and Sir John. This was a substantial company involved in large projects, and had established a mechanical engineering works at Blackfriars as early as 1810. Sir John headed the part that built canals and carried out naval civil engineering projects; George supervised the mechanical Engineering side of the family business, designing and building bridges, engines for ships, railways and associated rolling stock. In addition to their engineering works at Blackfriars they operated a marine engine works in Greenwich, London. There is some evidence to show that Thomas spent time in both branches of the Rennie business. At this time apprenticeships were normally of five years duration and usually required a premium. John Cowan would have paid this, and it is very likely that he was acquainted with the Rennies. Life as an apprentice usually started at the age of 15, and involved twelve-hour days.

Thomas would have attended the School of Mines up to two evenings a week. In many ways this is an exact parallel with a contemporary, Alfred Yarrow, (later Sir Alfred), and it is clear from a later entry that the two knew each other whilst both were apprentices although Alfred was younger than Thomas. Thomas was a member of the Civil and Mechanical Engineers' Society founded by Alfred Yarrow, and two of his talented friends, James Hilditch and Walter Rutt, in 1859, for the purpose of reading and discussing papers on engineering and allied subjects. There were initially about thirty members, the three founders being eighteen years of age or younger and all apprentices. Meetings were held in the evenings after work, and some well-known names occasionally presented papers. All the Engineering Institutions of today have similar junior sections.

In the 1861 (April), census John Cowan described himself as "fund holder", an indication that he had or was about to retire from the family business and pass it on to Thomas, despite this being widely reported as 1862.

An early record of T. W. Cowan's official engineering career is contained in patent number GB809, application date, 2nd April 1861, sealed 26th September 1861. "Improvements in the Means for Actuating Machine Hammers, which said Improvements are also applicable to Pile-driving and other such like Machines and Purposes". Application by John Grieve Winton and Thomas William Cowan, both of 42, Bridge Street, Blackfriars, London, indicating that on the 2nd April Thomas was still at Rennies in Blackfriars. The patent document comprises ten pages of text and four double page drawings, of great detail. John Grieve Winton was a Scottish Engineer at Rennies. This patent deals with the use of compressed air to aid steam hammers.

In his diary for 1864, Henry Michell wrote:

T W Cowan .... was now settled in business on his own account in Bridge St., Greenwich ....

It is typical of the confused information that surrounded Thomas; Bridge Street was not the home of Kent Iron Works, but the above shows how easy it is to be misled and how the misinformation is easily generated.

In 1862, at the Alexander Palace Great Exhibition, T. W. Cowan exhibited a machine that uses the principle in the above patent for small mechanically driven hammers. The entry in the official exhibition catalogue and comparison with the Patent documents, shows that the two are the same.

Essentially this is a basic belt driven hammer, with an added means of varying the blow without changing the weight of the hammer or the distance it drops. The number of blows in any given time was varied as normal by using conical pulleys. It would have application for a smith where no two successive jobs were the same, and did not require great power. However, the claims made for the range of operation are exaggerated, and whilst steam is not required for the machine itself, the probability is that in most situations it was, to supply the drive, and hence could easily be used for control also questioning the justification. It did not receive an award.

James Nasmyth had invented the steam hammer in 1837 and patented it in 1842. He quickly devised a means by which both speed and 'blow' could be varied, initially being controlled by hand gear. This was followed by a short period when self-acting gear was added, but it was found to be less effective than the original hand control, and was abandoned. In 1843 he admitted steam above the piston to aid gravitation, in the same year devising the steam hammer pile driver, and in 1848 the steam hammer form of steam engine. He demonstrated the versatility of his design by cracking an egg in a glass without breaking the glass, followed by a blow that shook the building they were in with the next. It was exhibited at the previous show of 1851.

William Herrod-Hempsall in *Bee-keeping Old and New,* p 1684/5 wrote of this machine:

```
...a compressed air hammer, the only one of its class (correct),
the others being all steam driven.  The fall of the hammer
(8-cwt.), could be regulated from 1-oz to 50-cwt., and
delivered 360 blows per minute.
```

CLASS VII. -*Manufacturing Machines and Tools.*

COWAN, THOMAS WILLIAM, *Kent Iron Works, Greenwich.*—Patent air compressed machine hammer for general forging.

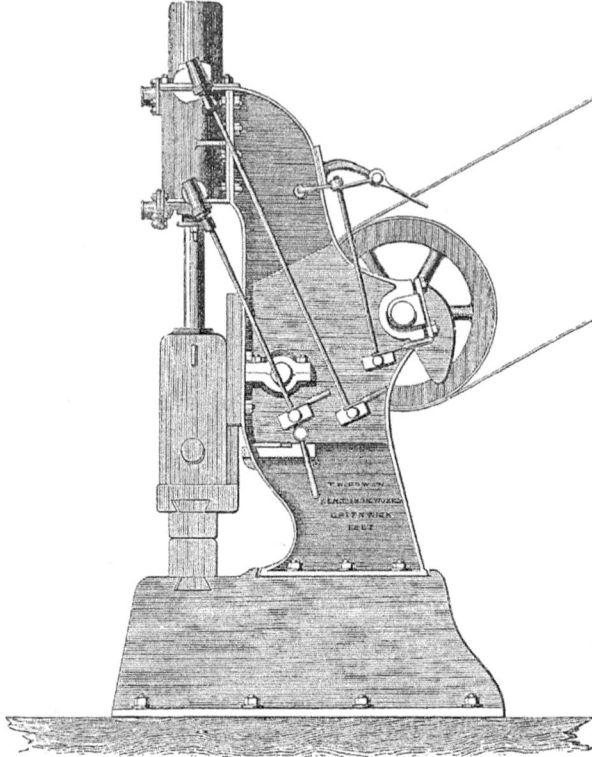

PATENT AIR COMPRESSED HAMMER.

The engraving represents an AIR HAMMER of 8 cwt. without compression. These hammers can be regulated to the utmost nicety for giving a blow equal in weight to the fraction of an ounce, and increasing it to about 45 cwt. The following is a description of the way this is effected :—

In the first place the motion to the hammer is transmitted by the strap and cam through a lever, which is raised every revolution of the cam. For a very light blow the cock at the bottom, which is usually open, is at this moment shut, causing the air to be compressed at the bottom of the cylinder, at the downward stroke of the hammer ; this blow can be regulated by partially opening the cock. When a blow is required equal to the weight of the hammer itself, all the cocks are open. When a heavier blow is required the air is compressed by regulating the upper cock, which communicates with several chambers.

These hammers are recommended for general smiths' work, as they are very easily managed by any boy. The hammer block can be suspended at any part of the stroke, and the speed may be regulated the same as steam hammers.

Small hammers in sets of twos and fours driven from one shaft, and having conical speed pulleys, are very useful where rapidity of workmanship is required, as they are capable of giving about 360 blows per minute.

Trunk hammers are made on this principle for drawing out steel, &c. ; also movable cylinder hammers for the same purpose.

All these hammers have very heavy anvil blocks, and the main frame being fixed on these blocks it is impossible for them to sink in the ground without the whole machine going together ; hence there is no danger of breaking any of the parts.

None of these hammers require massive foundations.

There is no expense in having to keep up a boiler with high-pressure steam for these hammers, as they are driven by a strap from the usual main shafting, and there is very little foundation required. They are perfectly under the control of the hammer-man, and very soon pay for themselves. Price, from £65 upwards.

WINTON & COWAN'S PATENT HIGH AND LOW PRESSURE DOUBLE CYLINDER HAMMERS are recommended for large forgings. These hammers are made to any size required, the smaller ones having single frames, as the drawing above, and the larger ones, for iron manufactories, double frames, they being best adapted for manufacturing iron and steel.

These hammers effect a great saving in steam, as the steam which is used in raising the hammer, after it has done its work in the small cylinder, is allowed to enter the large cylinder, and give the blow. Price, from £100 upwards.

(42)

Entry in the 1862 Exhibition catalogue

The final two paragraphs on "Winton and Cowan's High and Low Pressure Double Cylinder Hammers", is where patent GB809 is fully applied. In these hammers, steam is the only source of external energy, recycling it to reduce energy loss and improving performance as per Naysmyth. However, the Winton/Cowan patent also applied compressed air as in the previous mechanical hammer. The Kent Iron Works machine appears to be aimed at the small blacksmith where there was not steam on site.

Patents cost nearly £500 to submit and obtain at the time; not money well spent, I expect. However, it is not known if many, or any sales were obtained or whether Rennies or Kent Iron Works would have been the benefactor.

The second exhibit at the 1862 show was a patent nominal 16-horse power trunk engine, the patent being held by Burgh and Cowan. The trunk engine was one of many inventions by Watt, the original design being in 1784. It was very expensive to operate and there were problems because of the exposure of the trunk to ash and other substances, resulting in excessive friction at the trunk stuffing boxes. This problem was solved with the invention of the double trunk engine patented by a consortium involving John Penn and Maudsleys, both large engineering companies of London. I believe Burgh to have been an engineer with John Penn; he claims responsibility for the design.

In 1865 N. P. Burgh Esq. Engineer, read a paper to the Society of Arts on "Marine Engines from 1851 to the Present Time". It was reproduced in *English Mechanic*. Details of machines made by Dugeon; Blackwall; John Penn and Son; Greenwich; Maudsley, Sons and Field, and R. Napier and Sons, Glasgow, were given and he ended thus:

> Lastly I allude to the writers invention "Burgh and Cowan's Patent Antifriction Trunk Engine" so arranged that the friction of the trunks is dispensed with and no area lost in the cylinder. This arrangement was represented (at the 1862 exhibition), by a pair of engines and drawings.

So there is no doubt that Burgh was the inventor/designer and Cowan the manufacturer. However, the Patent Office has no record of the patent.

The Field mentioned above is Joshua Field, who was a partner in the company. He was a well-respected Civil Engineer, and one time President of the Institution of Civil Engineers. It was at his country residence that Fanny and Thomas were introduced.

CLASS VIII.—*Eastern and Western Annexes.*

COWAN, THOMAS WILLIAM, *Kent Iron Works, Greenwich.*—Patent nominal 16-horse power trunk engine.

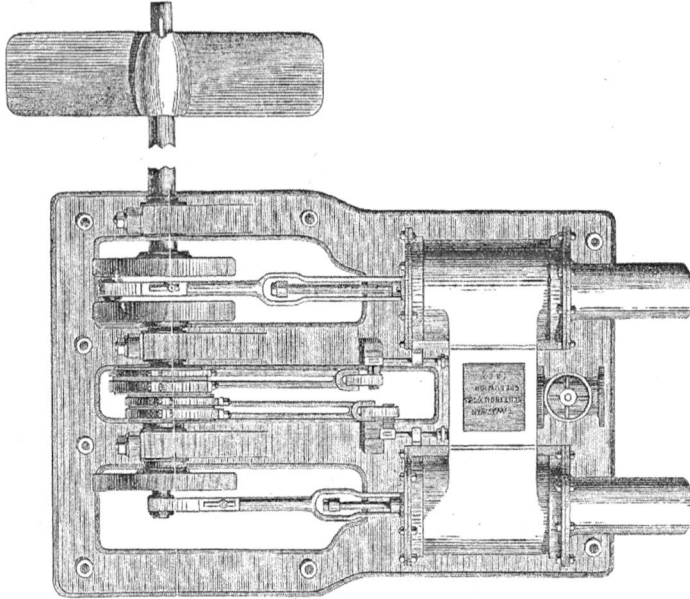

PATENT TRUNK ENGINE.

BURGH & COWAN'S PATENT TRUNK ENGINE of 16-horse power.

It is a well-known fact that the trunk engine is the most simple at present in use ; but the immense friction of the trunks in their respective stuffing boxes, and their alternate exposure to the steam and atmosphere, render them highly destructive to steam and tallow.

T. W. Cowan, the sole manufacturer of Burgh & Cowan's patent engines, is desirous of introducing them to the public. The following are a few of the many advantages gained by the use of these improvements :—

1. The area is gained, hitherto lost in trunk engines, thereby a saving in space.

2. The immense stuffing boxes being entirely dispensed with, a great reduction in friction and packing is effected.

3. The trunks are never alternately exposed to the steam and atmosphere ; also the moving or piston trunk is entirely frictionless, gaining a considerable saving in tallow and in steam.

4. The connecting rod is in the centre of the cylinder, and perfectly accessible to tighten and lubricate, which dispenses with the guides beyond the cylinder.

5. The guides being within the cylinder, and cast in the trunk, they never get loose, and are entirely out of harm's way.

6. In beam engines, this improvement entirely dispenses with the expensive and complicated parallel motion, thus rendering engines cheaper and simpler than those at present in use.

7. In high and low pressure engines, the high pressure is within the low pressure, while the areas of both are maintained. This is a great advantage over those at present in use.

8. The simplicity of the whole engine, together with the small space it occupies at any given horse-power, renders it highly advantageous, particularly for marine purposes.

9. In stationary engines the connecting rods are about six times the length of crank.

10. Marine engines made on the same principle as the above engraving are much lighter, take up less space, and are much cheaper to work than any other description of engines.

11. Vertical engines on this principle are particularly adapted for places where there is little room to spare ; a 10-horse power engine only taking up the space of 1 foot 4 inches by 1 foot 8 inches.

12. In steam fire engines the pumps are connected by a rod, to the piston, through the bottom trunk, thereby taking up less space.

13. The high and low pressure engines are invaluable where fuel is expensive, as they save a great deal of steam that is altogether lost in other engines.

These engines being of the best materials and workmanship are found to be cheaper and work longer than any other engines.

CLASS VIII.   (17)

Exhibition documentation for trunk engine

The claims made for it are extensive and difficult to appreciate e.g. 'the moving piston is entirely frictionless', and 'the high pressure is within the low pressure, while the areas of both are maintained', but it was, as before, a promotional document.

The plate on the engine reads "T W Cowan Kent Iron Works Greenwich 1862."

The third exhibition entry involving Cowan was a steam carriage for use on the road. Again to quote Herrod-Hempsall:

> ....were exhibited three examples of his skill...... The most interesting was a steam carriage, one of the first used on English roads to carry 12 passengers, and capable of travelling at 20 miles an hour. ...we present ...a photograph of this carriage on the road, with Mr Cowan at the wheel. The carriage was improved to do 30 miles without a stop.....

Trevithic, the well-known Cornish steam engine designer and manufacturer, built the first British road steam carriage in 1801. The following year the Felton Carriage Co. of Leather Lane, London, built one of Trevithic's designs for a steam road car and drove it 10 miles through the streets of London with 7/8 guest passengers. Unfortunately it crashed and he gave up the idea as not a realistic proposition, mainly due to the volume of fuel and water needed to be carried to travel even small distances. It also suffered from limited ability to manoeuvre. However, in 1834, the Steam Carriage Co. of Scotland built six vehicles that carried up to forty passengers at speeds of up to 20 m. p. h. and two of these eventually operated in London. So the idea was certainly not as ground breaking as Herrod-Hempsall would have us believe.

The catalogue for the 1862 exhibition does not have an entry attributed to T. W. Cowan or the Kent Iron Works for a steam road carriage. However it does have one attributed to Alfred Yarrow and James Hilditch with an attached drawing as shown. Cowan is at the wheel and Alfred Yarrow is one of the passengers.

Due to an unfortunate series of mix-ups the carriage was not judged, and therefore received no award. The judges later sent Yarrow and Hilditch an apology for the error, but not Cowan.

The following extract is taken from *Alfred Yarrow his Life and Work* and was accompanied by the plate reproduced.

> It appeared to Yarrow that a field was open for mechanical propulsion of vehicles on an improved plan, which should render them reliable and simple to work; and with Hilditch he designed and patented a special type of steam carriage

in the latter part of 1861. This invention was taken up by T, W, Cowan, of Greenwich, subject to payment of a royalty.

A carriage was built and driven from Greenwich to Bromley, a distance of 10 miles, once a week in late evening. At Bromley, Yarrow and some possible purchasers would stop a short time to take some refreshment, and then return to Greenwich, catching the midnight train to London. The swift passage of this steam carriage at night created a good deal of interest and excitement. One old lady, who heard it rushing past her house, ran to the window and, seeing flames issue from the funnel of the boiler, declared that the devil was at hand. Unfortunately, on one occasion, the carriage, when going at about 25 miles an hour, met a mounted policeman. The horse took fright at this surprising apparition, and the policeman was thrown to the ground, breaking his leg. This, among other causes, led to a bill being passed forbidding the use of steam upon the road, unless the carriage were preceded by a man carrying a flag. This put an end to the nocturnal trips, and to the evolution of the motor-car till the law was altered.

So the combination of bad luck and legislative interference meant that this business could not be looked upon as successful for any of the principles.

At the 1862 exhibition there were 29,000 exhibitors, 9,000 from Britain, and 40% of exhibitors received awards; Cowan and Kent Iron Works were not amongst them.

It is clear from this that Cowan was yet again, simply the facilitator of Yarrow and Hilditch's design. At the time the two designers were 19 years old and the carriage was an improvement on those that went before in that it had improved suspension and greater manoeuvrability.

The plate on the side of the steam carriage reads 'T W Cowan Marine Portable Engine Maker Kent Iron Works Greenwich 1862'. This appears to indicate that his business was now concentrating on making small marine engines, to the Burgh and Cowan patent.

Thomas 'resurrected' the steam carriage in the *English Mechanic's* issues of April 30th and May 14th 1869. The engraving was taken from Yarrow's photograph; Kent Iron Works plate removed and the background redrawn, indicating that the business no longer existed. The stoker still looks overdressed for his part in the operation. His description has been copied directly from the magazine and is given below the engraving.

Steam carriage designed by Yarrow and Hilditch, made by Cowan

ENGLISH MECHANIC

AND MIRROR OF SCIENCE

Engineering, Building, Inventions, Electricity, Photography, Chemistry, &c.

VOL. IX.—No. 214.   FRIDAY, APRIL 30, 1869.   [PRICE TWOPENCE.

COWAN'S IMPROVEMENTS IN STEAM CARRIAGES ON COMMON ROADS.
NO. 6, VOL. IX.

STEAM CARRIAGES ON COMMON ROADS.

SIR,—According to request 1 send you a description of the steam carriages constructed by me. The first carriage of which you have a photograph is of 4-horse-power, the total weight of which is 25cwt. The framing which carries the engines, boiler, and carriage is made of ash, covered on the outside with a 3-16 in. plate, and is 13ft. 4in. long, and 4ft. wide. On each side of this frame cast iron plates are fixed carrying' cylinders 5in. diameter. The two driving wheels which are 3ft. in diameter and 3in. broad, are placed inside the framing, and on each end of the shaft are keyed the cranks. The wheels are iron and wood (Crosskill's Patent). Between the wheels and the cranks are the bearings which are on springs attached to the main frame. To maintain the proper distance between the shaft and cylinders, instead of axle-box guides, as in locomotives, a link is adopted connected at one end to the bearing-block, and at the other to the engine-plate, thereby reducing the friction to a minimum. By this means the valve gear is also unaffected by the motion of the springs and the engines consequently much simplified. The engines being placed on the outside of the framing, and connected direct with the driving axle - no chain or cog wheels are used, and they work with great ease and smoothness. The link motion and valve-gear are between the two cylinders, and the eccentrics are keyed on to the shaft between the wheels. The brackets, &c, carrying the valve-gear are all of malleable cast iron, and the valve gear itself of steel. There is a great advantage in placing the driving-wheels inside the framing, as by that means the cranks can be keyed on to the ends of the shaft, instead of using a double throw crank shaft, which at a high speed is very liable to bend and throw everything out of gear; also, the wheels being closer together, can, therefore, both be keyed on to the shaft, and thus the adhesion of both wheels is obtained without the complication of clutches to throw them in and out of gear when turning- curves. The engines are completely covered in to keep them from the dust. We now come to the boiler which is fixed at the back end of the carriage. It is upright tubular, of steel 3-16in. thick, 2ft. diameter, and 3ft. high. It is constructed to work, if required, at a pressure of 150lb. per square inch, and is tested to 300lb. Below the boiler is an ash-pan, sloping down towards the front of the carriage, and has a door fixed which is raised or lowered to regulate the draught. At the back of the boiler is the tender where the stoker stands, and which contains the coals. The fore carriage is similar

in every way to that of an omnibus, except that instead of
being turned by a pole, as in a omnibus, it is guided by a
chain passing over a worm fixed to the steering-wheel placed
in front of the steersman. It will be seen that the seats
are arranged in the form of a wagonette, to carry twelve
persons, the steersman sitting in the centre of the front the
seat, steering wheel in front of him, a little to the right
the reversing lever for controlling the speed, reversing the
engines and stopping them. The break is worked by the left
foot, and is in connection with both the driving wheels.
A steam-gauge and a whistle are also fixed in front of the
carriage. Besides the entire control the steersman has over
the engines, in case of accident, the stoker can also stop
them and put on the break from his end of the carriage. The
boiler is provided with two spring balance safety valves a
steam gauge, and steam whistle. The tanks are under the
seats, and the boiler is fed by a small pump when the engines
are at work, and by an injector when standing still. A steam
jet is introduced into the funnel to assist in raising
steam when going up steep incline. When this carriage was
first built, the valve gear was of wrought iron, and was
constantly twisting, but when the steel was substituted,
it was rendered perfect. This carriage has no condenser,
and the waste steam, consequently, rushes up the funnel,
creating a great draught, and entirely consuming the smoke.
The carriage is very easily steered, and can be stopped
within a couple of yards. The speed on a level road with 100
lbs pressure (the usual working pressure) is about twenty
miles an hour. The highest speed I have attained is 30 but
this I consider dangerous, as should-anything happen to the
steering chain, before the engines could be reversed, the
result would be most disastrous. It will ascend a gradient
of one 1 in 16 with ease, and the steepest gradient I have
ascended is 1 in 12 at the rate of about 5 miles an hour. On
an ordinary road the speed is about 15 to 18 miles an hour.
In the neighbourhood of Bromley, in Kent, I have run 6 miles
in 22 minutes. Travelling in it is quite a pleasure, being as
comfortable and as easy as in a first-rate brougham, and the
ease with which it is steered and guided even at the highest
speed is the astonishment of all. I have also constructed a
second steam carriage, in every respect similar to the first,
with 6 ¾ in. cylinders, as surface condenser, and, of course,

all the other parts proportionately larger. Its weight was 2½ tons. The condensor was only used when passing horses in the country. Coke was used, so that there was no smoke or noise from the steam. I may mention that the link J keying the wheels inside the framing onto the axle, are Yarrow's Patent. This class of carriage is evidently adapted to work in conjunction with railways from stations to neighbouring towns, where the traffic is not found sufficient to compensate for the outlay of a permanent way; in any case where the stagecoach or other old-fashioned conveyance is in use. The adoption of these carriages would be found to result in great economy. I t is a great pity they are not allowed to run in the streets of London, as they are certainly not so objectionable as trains running over the ugly bridges in all parts of the metropolis. 1 may here mention that the reason horses are frightened at traction and other engines, is, because they have blinkers and cannot see on each side of them. Abroad, where these ornaments are not used you seldom hear of horses being frightened and running away. In conclusion, I may say, I shall be happy to furnish any further particulars that may be required.
T. W. COWAN.

T. W. Cowan was still smarting at the failure of the carriage. It was modified for use in farming as the motive power for ploughing, again without success. When using mechanical power rather than horses, ploughing utilised stationary steam engines which pulled the plough across the field with cables; relatively unattractive compared to using the horse.

There are two further patents attributed to T. W. Cowan. GB2603. "Improvements in the Construction of Breech-loading Ordinance". Application date 18th October 1861, not sealed. This consisted of an improved form of breech-piece, which he proposed applying to all kinds of guns. The description lacks detail as is normal at the application stage of a patent, but the 'improvement' was a rotating bridge piece with two or more chambers – a large scale 'six shooter' springs to mind. It was never sealed, indicating that it was not a new idea, and accompanied with considerable difficulties in 1862. In the preamble Cowan wrote:

I, Thomas William Cowan, of the Kent Iron Works, Greenwich, County of Kent, Engineer, ...

Thus confirming that it was 1861 that he took over the family business from his father.

FIGURE 1.

FIGURE 2.

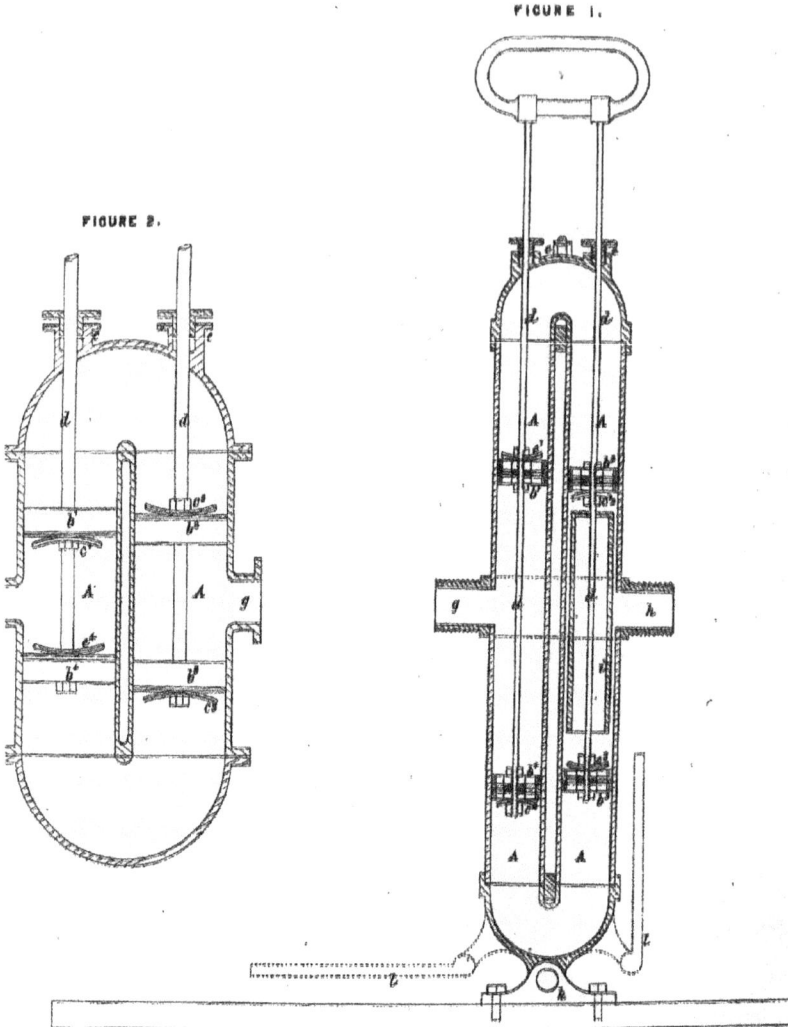

be filed drawing is not colored.

Drawn on Stone by Malby & Sons

LONDON: Printed by George Edward Eyre and William Spottiswoode, Printers to the Queen's most Excellent Majesty. 1863.

Diagrams for Pump Patent

Patent No. 2525, application 13th September 1862, sealed 6th March 1863; "Improvements in the Construction of Portable or Fixed Pumps". The improvement lay in the construction of force and other pumps by means of the introduction of four pistons or valves with a double-barrelled pump. In the patent he described it thus:

The apparatus consists of two barrels parallel to each other, each having a double action within itself. On the

piston or valve rod in one barrel the valve acts so as to
draw the water or other liquid in. the leather or other
material forming the valve in the piston being placed on
the upper side of it; the other and lower piston in the same
barrel having the valve or flap on its under side.   This
arrangement is reversed in the other barrel, the upper
piston having the valve on its underside, and the lower one
having the valve on its upper face. By this arrangement
two pistons will be sucking or drawing in the water or
other liquid, and the other two forcing it out. On one
of the valve rods I place an air vessel, which moves with
the valve rod and within the barrel. This form of pump is
applicable to any sort of work, such as fire engine, garden,
feed or other pumps.

He appended the diagrams shown. The link between the trunk engine and
this pump can be clearly seen.

There is no evidence of its success or otherwise, but Thomas was selling
them after the Company ceased trading. The local Industrial Archaeological
Society, always believed that Kent Iron Works were situated adjacent to the
Merryweather fire engine business, which was in Greenwich High Road. In
the early 1860s the company was called Merryweather and Field. Edward Field
was consultant on the boiler part of the business. His name crops up again!
However, I have no firm evidence that they installed Cowan's pumps.

Herrod-Hempsall also claimed that:

Mr Cowan was an expert on sewage and drainage, and held
the appointment of engineer for several schemes. In 1870
when a channel tunnel was mooted, he advocated the use of
a sectional metal cylinder for its construction….

I have been unable to verify these claims. For many of the small schemes
no such records survive, and his tunnel proposition was just that, and not
new, it had already been used in London and featured in *English Mechanic*.
Henry Maudsley who had a works at Lambeth, had been associated with John
Rennie, and had produced the shield for the Thames tunnel, the Worlds' first
underwater tunnel. I believe that Herrod-Hempsall is writing about projects
that Cowan was involved with whilst an apprentice at Rennies. It is typical of
the upgrading of information from both Cowan and W. Herrod-Hempsall.

Following his marriage to Fanny and Henry Michell's gift of £1000 on his
wedding day, Thomas bought "some mining business" in Wales. Evidence
shows this to have been the Tryfanant mine. In his diaries for 1864 Henry

Michell wrote:

> ...but Mr Cowan had during the summer in short very soon after his marriage taken some mining business at Trefynant near Llangollen in Wales for at the end of October we went down there with them to look at this new concern which I was solicited to join but I could not see my way clear enough to commit myself to unlimited liabilities I did tender my son in law some assistance in money and of course hoped it would turn out well but must say I hoped against hope.
>
> Oct 24th My wife and myself went to Llangollen and returned on the 27th. We went partly to see Mr Cowans business at Trefynant was progressing I of course knew little of such matters but it did not seem to promise very well but we hoped all would end well upon the whole we enjoyed our trip very well...

The Tryfanynt coal mine was a small drift affair, employing about three hands. It is difficult to understand why he said that he "knew little of such matters" because he had, at one time in his business life, been involved with the 'coal trade'. However, it is clear that he was not impressed with his son-in-law's purchase. In the *English Mechanic* of January 23rd 1869, Cowan wrote:

THE THAMES EMBANKMENT.
Sir,—In a leading article of the " Times," of the 21st ult., referring to the financial embarassments of the Metropolitan Board of Works, it was stated, "We have no reason for supposing that the Board has been recklessly extravagant. So far as we can tell the outlay on past works was not excessive, &c." And before the end of the week the same assertion was repeated. I therefore think it well to give you a statement of facts, and leave it to your readers to judge how far the assertion is correct Unfortunately we have very good reason for supposing that the Board has been recklessly extravagant, and to prove the truth of my remarks, I shall take the Thames Embankment as an instance. In 1861 a Royal Commission was appointed to make inquiries as to the resources of the country in respect of stone, available for public buildings in London and elsewhere.

The Report of the Commission was published, and at p. 55, in his late examination, the late Mr. J. R. Burnell stated that he had been sent that week to examine in the Vale of Llangollen the millstone grit, known by the name of Garth stone.

Subsequently, Mr. Tite, as a member of the Metropolitan Board of Works, brought this game silicious grit stone under the notice of the Board, as suitable for the Thames Embankment, being fully equal, if not superior, to the granite previously used; and, owing to the facility with which it could be cut or split, could be worked with considerably less than half cost.

Mr. Burnell also stated he had visited the Vale of Llangollen, and had examined the stone, had ascertained the varieties and thickness of the various beds, and had examined the ancient structures in which this stone had been used, among which are the Castle of Dinas Bran, probably the oldest ruin in the kingdom, the Abbey of Vale Crucis, and the Bridge of Llangollan. He likewise stated that, not only the ordinary exposure to the air, fire, and water, for a period of seven or eight centuries, had exercised no perceptible influence upon the stone, but that even the buttresses of the Llangollen Bridge, exposed to a mountain current, alternating constantly between the strongest stream and long periods of dryness, still exhibited the marks of the workmens' tools seven centuries ago, and thus established, to the best of all possible tests, its durability and strength. He lastly stated that it could be delivered in London at less than half the cost of the granite.

Also the mountain limestone was proposed on account of its cheapness and durability; but, in 1865, the Board decided that, as it was found that the use of the mountain limestone would only produce an economy of about £16,000 or £17,000 in comparison with the granite, it was not worth while to employ the cheaper material for that purpose, and therefore granite was to be employed in the execution of the works. Previously to coming to this conclusion, the Board had disposed of all the claims of the millstone grit, on the score of the great absorption of the stone, which decision was founded on a hasty and prejudiced view of the matter.

It may not be worth noticing that the saving in this case

would have been only about £16,000 or £17,000 (by their
own stating, although in reality it would have been nearly
double, as the limestone or millstone grit, are both easier
worked than granite), yet it is worth mentioning that as
one material is equal to the other in durability, it would
have been a considerable and advantageous saving. I think
the Metropolitan Board are bound to execute the works they
undertake at the lowest possible cost consistently with
the good, sound execution of them, and I think any other
principle leads to a degree of reckless extravagance so
painfully visible in the Government works of this country;
and I confess I am not all surprised that the Board are in
any financial difficulty, especially as they so ostentatiously
displayed their disregard to what might have proved in the
end a very great measure of economy.

A committee was appointed to report upon the various stones
submitted for the execution of the works, and the attention
of the gentlemen was especially directed to only three kinds—
the millstone grit, the mountain limestone, and the granite—
but nothing was said of the particular kind of granite,
although it varies very much in its character, according
to the locality it is procured from. Their objections were
certainly not valid. The millstone grits, they stated, were
too absorbent, and they excluded them on that account; the
mountain limestone was objected to because it was attackable
by the saw and sand, and would yield (they said) to the
influence of weather, or be acted upon by the acids in the
Thames water.

On the validity of these objections much might be said, but
I do not think that the advantages on the side of granite
are so great as to have excluded its competitors from the
market.

With regard to the millstone grit, it is a fact it does absorb
water very rapidly, but it can have very little effect upon
the materials of which it is composed, as the silica exists
in the form of distinct crystallisation; and even in the
Garth stone the cementing material is silicious—there is,
therefore, no room for the action of the atmosphere. As a
proof of the invalidity of their objections to the millstone
grits, I can mention the piers of Southwark Bridge, which

have stood the tidal influence of the Thames, thereby proving the adaptability of this stone to the very purposes the Board required. The bridge at Llangollen, has stood the effects of exposure to the action of the atmosphere and water for now more than 500 years, yet the Dee rolls down, as I myself have witnessed, rocks, trees, &c, with its winter floods, that are much greater than the Thames ever transports in the tidal part of its course.

The decision by which the mountain limestone was rejected, accompanied by Professor Tennant's report, was the most unwarrantable piece of proceeding that has ever been witnessed.

The Professor first declared the samples were picked, and then proposed that they should be subjected to the action of the tides for two or three months, in some position where he could examine them. He, however, forgot to mention how this stone had stood the tidal action in the Menai and the Britannia bridges, and the experience to be derived from a trial of half a century must be more conclusive than that of two or three months which the Professor himself proposed.

The committee did not take the trouble to visit any of these works, and were content with the objections of Professor Tennant.

This stone was also rejected, and the conduct of the committee appeared to have been influenced by a preconceived opinion, and apparently they were led by that opinion to disregard the economy that the limestone would have introduced into the work. This stone exists largely in England, Scotland, and Ireland; and in Belgium it is universally used in all cases where it is desired to unite strength and durability— in fact, the Belgium quarries have the complete monopoly of all the docks, river, and canal works.

Yet the Board refused to use this stone, and recommended the use of granite in its stead, and at a time when they were under difficulties with the contractors, who stated they were not able to procure material in sufficient quantities for the northern part of the embankment. The price of labour on the mountain limestone must be at least half that of granite;

this, however, was not taken into the account of saving, but it would have greatly added to the net sum economised.

The conduct of the committee in this particular case has been marked by the extravagance which so often attends the conduct of those who work with the money that is raised from the public, and towards which they themselves contribute a very small fraction. It was not an economy of £16,000, although that would have been a large sum, especially in the present state, but perhaps a question of considerably over £100,000, if they embank the south side of the Thames.

As payers of the rates, we ought to bear all this in mind, and I think, after what I have stated, every one will agree with me that the Board in this particular instance was recklessly extravagant.

I could go into the merits of the various stones more fully, and show that the Thames water would in no way, chemically or otherwise, affect the millstone grit or limestone, but feel I have already trespassed too much upon the space of your valuable journal.
THOMAS WILLIAM COWAN, F.S.A., &c.

I have reproduced the complete letter because I believe it to be the first published item that shows parts of his character that are less than attractive. There are many instances in the future, when the same pattern emerges: - he starts by giving his version of the facts "so that readers can judge for themselves", quickly adding "everyone will agree with me that…" or something even more insulting.

The North side of the Thames embankment cost £1,599,055 and the South £946,010, according to a later answer to correspondent in he same magazine. But why was Cowan so irate at the decision? Could he have acquired a quarry business to add to the coal mine and anticipated a considerable financial gain, or did he think that if Garth stone were chosen, his mining business there would benefit indirectly? The letter has all the signs of a vested interest.

The accompanying document to a vase, design register number 198204 in 1866, gives Cowan as the proprietor of Trefynant colliery, confirming that it was a coal-mine that he owned, and still did in 1866. The notes give the impression that it was a ceramic vase.

The Cowan vase

Some relevant dates in respect of this particular chapter: -

1861 Owned Kent Iron Works. Most writers have this as 1862, but in the 1861 census John is listed as a fund holder and Thomas a Civil Engineer, a patent entry confirms. The Yarrow biography gives the same date.

1862 Exhibited at Great Exhibition, catalogued as Kent Iron Works.

1864 Married. Purchased 'mining interest' in Wales.

1869. Moved to Horsham.

1871 Census, Thomas listed as civil Engineer, Ph. D., M. A. etc

1874 Henry Michell died.

1875 Henry Michell will proved. Both Thomas and Fanny benefactors.

1878 Built (occupied), Comptons Lea

1881 Census, listed as farmer of 40 acres.

1886. Thomas inheritance from Henry Michell ended. Inherited from own parents.

1889/90. Left Comptons Lea.

1891 Census, "on own means".

Because Kent Iron Works was not a limited company, there are no records in the National Archives and much research effort by local industrial historians has failed to produce any positive evidence. However, I believe that Thomas's association with Kent Iron Works ended in 1869, and that the business ceased trading at the same time. The evidence for this deduction is: -

1. The Cowan family moved from Beckenham to Horsham in 1869, and the entry in Henry Michell's diary, quoted earlier.
2. Thomas advertised several items made by his engineering business or used in it at this time, in the *English Mechanic;* might also indicate a lack of money.
3. The plate of Kent Iron Works did not appear in the engraving of the steam carriage in the *English Mechanic.*
4. In a trade directory covering Greenwich for1870, there is no mention of Kent Iron Works.

However, in the 1871 census Thomas still lists himself as a civil engineer; possibly his role with the Henry Michell businesses.

In conclusion; from 1869 Thomas worked in the Michell family business. In 1873 Henry made him a beneficiary in the first Codicil of his Will but only until 1886 when coincidently he inherited from his own parents. A few years later he sold Comptons Lea and moved to London, indicating that his presence in Horsham was no longer necessary, and his association with his wife's family business had ceased, probably when his money stopped.

I received considerable technical assistance from an old Mechanical Engineering colleague in the preparation of this chapter, but he wished not to be acknowledged.

Chapter IV

Early Bee-keeping Activities

It is believed that Thomas William Cowan's interest in the Natural World was first aroused as a schoolboy in Russia and developed when he was an occasional student at *The London School of Mines* in his late teens. That Institution was adjacent to, and had an intimate association with The Museum of Economic Geology where he could have attended lectures by Charles Darwin, Thomas Henry Huxley etc. By 1860 he had become very interested in the scientific aspects of the honey-bee and it's domestication, communicating with Alfred Neighbour, a merchant, and Thomas White Woodbury who was known as 'The Devonshire Bee-keeper', and a prolific writer on bee-keeping in the *Journal of Horticulture and Cottage Gardener.* Woodbury was considered one of the leading bee-keepers and authors on bee matters of the period, despite never publishing a book. He was one of those individuals that Darwin communicated with regarding honey-bees, especially the difficulty he had with their altruistic character. Despite his earlier inclination towards engineering, influenced by his father's profession, study of the Natural World in general and honey-bees in particular, would be continuing themes for the remainder of Cowan's life.

T. W. Cowan first kept honey-bees in 1864 when he resided in Rhyde House, Beckenham, his first marital home with Fanny. Beckenham had easy rail access to Greenwich and the Kent Iron Works, Thomas's place of work. By the 1873 bee-keeping season, when the Cowan family had moved to South Street, Horsham, Thomas reported that he had one Stewerton hive, one colony in a circular wooden construction, one in a Neighbour box, and nine Woodbury hives with either ten or thirteen frames, housed in his apiary above his stables. He later corrected this to indicate that he had modified the Stewerton and Woodbury hives. The frames he used did not have a bottom bar, but he claimed that the bees did not build comb down and attach it to the floorboard. He appears not to have experienced a problem keeping the frames square. At this stage he did not detail the modifications he had made to the original hives, but later informed that he had changed the Woodbury hives to take either ten or thirteen frames.

In his first *British Bee Journal* article on ligurising (replacing native queens with Italian stock), he stated that his apiary comprised Stewerton and Woodbury hives, and whilst this is not a total contradiction to the statement in the previous paragraph, it reflects the difficulty writing about an individual for whom accurate, consistent information is often not available – even from his own pen.

Ligurising had been a regular topic in *The Journal of Horticulture and Cottage Gardener* some ten years before the appearance of the *British Bee Journal* and Abbott had written about it in the *English Mechanic and World of Science* in the early 1870s. Cowan wrote detailing a slight modification to the method in respect of queen introduction, but admitted that he had lost queens using it; there was little new.

When presenting a paper twenty plus years later, he reflected that he had practised storyfying (piling of boxes vertically), since 1864, and the supers that he exhibited ten years after this were in Stewerton boxes and worked on the top of frame hives. He had used comb foundation from the beginning, but remarks that full sheets could not be utilised then because they lacked strength. He bought imported foundation from Neighbour of Roots manufacture, and "Scotch sheets", probably from Raitt. He also had a pair of plates to make his own foundation but does not indicate success or otherwise or detail the method he used. He would most definitely have tried.

Cowan did not use additions to frames to space them, thus enabling the distance between frames to be varied, something he passionately believed in. He sometimes used spacing as small as 1¼" between centres of frames to prevent drone breeding and swarming, using his finger to set the distance. In the discussion following a paper on one occasion, frame spacing was mentioned, general approval being expressed with some of the devices available. Cowan was not present during this discussion, but afterwards expressed surprise that no one mentioned his method because he had written of it many times in the *British Bee Journal*. Perhaps they felt that if he wished his opinions to be contributed to the discussion, he should have been there; or they might just not have being paying attention!

Abbott claimed, correctly, that when the standard frame was decided upon in 1882, the sub-committee did not discuss the measurement or method of spacing frames. He had developed a distancing addition to his frames, and was not pleased that no recommendation was made on frame spacing. Cowan's continuous recommendation to vary the spacing of frames contributed to the animosity between them, and did nothing for harmony within the B. B. K. A.

The Rev. John Cumming, styled 'the Times Bee-master', writing in *The Times* on his 'take' for the year, brought forth Cowan's second article in the *B. B. J.* detailing his far superior honey yield, and casting doubt upon Cumming's bee-keeping abilities; a similar letter appeared in *The Times*. It was not uncommon for the 'Times Bee-master' to be the subject of abuse by correspondents of the *Journal of Horticulture*. Woodbury and Tegetmeir had a long running series of disagreements with him. Examination of his book, *Bee-keeping,* published in

1864, confirms their view that he was not a very talented apiarist.

Whilst in themselves the two initial offerings from Cowan were not important articles, they provide an early insight into Cowan's character

- I have improved a method of carrying out a well publicised bee management technique,
- I have developed a system of bee management that produces results far superior to those of others.

For someone who had only kept bees for a few years these were substantial claims, but common practice for the period, and was a continuation of his arrogant attitude exhibited earlier in *The English Mechanic and World of Science*. He was requested, challenged really, to give details of the system of management that had produced such impressive results. His response formed the substance of his third major article two months later, and showed that he had carried out some investigation and trials, but the time scale was very short and only involved a few colonies. Readers of the *B. B. J.* could now be in no doubt that a young knowledgeable but arrogant bee-keeper and writer was now amongst them. However, reading some of the efforts from his peers, would indicate that it would not have been difficult to impress them!

Cowan did not contribute to the debate taking place in the early issues of the *B. B. J.* on the wish for a bee-keeping organisation, and the desire of many to have a show as a catalyst to launch it. It was recognised that to run a show money was required. Early 1874 in the *B. B. J.*, Abbott opened a list for subscribers to the prize fund for the proposed show, also informing that a venue, the Crystal Palace, had been secured for the show in early September of that year, subject to adequate donations being pledged. The first list of subscribers was published in the May issue of the *B. B. J.* accompanied by an announcement of a meeting on 16th May to inaugurate an Association for apiculture for "all those who are subscribers to the prize fund". Cowan did not feature on the first list of eighty names.

At the meeting on the 16th May, 1874, chaired by the Hon. and Rev. Henry Bligh of Nettlebed, Oxford, the elected officers of the now named British Beekeepers Association (B. B. K. A.), were: -

President – not appointed, later Sir John Lubbock agreed to take the
        post.
        Vice Presidents – Hon. & Rev. Bligh, R. Symington
Treasurer – C. N. Abbott
        Secretary – J. Hunter
Committee – Messrs. C. Atlee, W. Abbott, F. Cheshire, J. M. Hooker,
        J. Turner, treasurer and secretary, who were to hold office

until the first A. G. M. of the Association.

This was effectively formalising what had been happening under the direction of C. N. Abbott. It was a very necessary step because there was now money involved. Four of those listed were members of the Abbott family.

The *British Bee Journals* for June, July, August and September 1874, contained lists of the contributors to the Crystal Palace Prize Fund, but Cowan was not amongst the number that eventually totalled over one hundred. The first list of members of the B. B. K. A. appeared in the August *B. B. J.*, again Cowan was not amongst the ninety seven names listed. However, he is amongst the one hundred and twenty names in the September issue. Hence in the period between 16th May and 1st September, Cowan had become a member of the B. B. K. A., but does not appear to have contributed to the Crystal Palace Show Fund which I find difficult to believe, but he did not request its correction.

At the first A. G. M. of the B. B. K. A. on the 10th September held in conjunction with the Crystal Palace show, chaired again by Bligh, Cowan was one of twenty eight people listed to be asked to be Vice Presidents, Lubbock being confirmed as President. The acting committee was to be Atlee, Cheshire, Smith Turner, (previously listed as Turner), E. Melladew, J. Hooker, W. Abbott, and T. W. Cowan, given in the order as printed in the *B. B. J.* The President, Vice Presidents, treasurer and secretary, were to be the general committee. In this very strange arrangement the responsibilities of the two bodies or the relationship between them is not defined.

This very detailed information shows that Cowan went from someone who apparently, was not interested in the formation of a formal organisation for bee-keeping, to becoming a member and at the first A. G. M. being one of two individuals added to the original core of committee members to form the new acting committee. All in a matter of months and apparently without making the requisite donation.

The Bee and Honey show at the Crystal Palace duly took place on 8th, 9th, and 10th of September 1874 almost a complete year after C. N. Abbot, F. Cheshire and R. Symington started their work towards it. In the interim, the Manchester Bee and Honey Show had been staged causing considerable disquiet within the bee-keeping fraternity. The trio were supplemented with others, a schedule, drawn up, the prize money raised and all other arrangements put in place. For a group of people that had done nothing even approaching this in the past, and with no template from which to work, it was a significant achievement. No comprehensive list of exhibiters survive, and might never have existed, but the schedule and successful entrants were fully reported in both the *B. B. J.* and the *Journal of Horticulture*, and in some instances other entries were also mentioned, especially when the reporter had difficulty coming to terms with the judges decision

My subject is first mentioned in conjunction with Class 13 - *For the best wood super of honey (or wood in combination with glass or straw) net contents above 20-lbs.* Writing in the *B. B. J.*, C. N. Abbott reported:

> This class seems to belong to another era, and here the competition was remarkable, there being no less than twenty-eight entries, the supers varying in weight from 28 lbs to 76 ½ lbs net. The first prize was awarded to W. H. Clark, Esq., of Morton Abbey, Surrey; the second to Mr J. Anderson of Ayrshire; and the third to Alfred Rushbridge Esq., of Chichester. Neither of these supers approached the weight, nor, excepting the second case, the beauty of the magnificent supers exhibited by T. W. Cowan Esq. of Horsham, whose description of them in the *Times* (and *B. B. J.*), a short time since quite startled the bee-keeping world, and many were the shrugs and winks implying doubt of their existence, yet here they veritably were, in all their pride and beauty; but the judges disqualified them because they could not taste them!

There appears to have been a considerable disquiet over this matter. Bligh in his usual conciliatory manner writing that it was an accident and no one was to blame, Hooker vociferously responded, outraged at the injustice that had been perpetrated:

> I think the judges exceeded their legitimate powers in disqualifying this splendid collection of supers, and evidence of skill on the part of the exhibitor.

In a later edition of the *B. B. J.*, editor Abbott responded to an inquirer – W. A., that Cowan's disqualification:

> ...was an arbitrary proceeding, and a great hardship on Mr. Cowan, since no intimation of the necessity for the honey being exposed had appeared in the Schedule. Some nicer means of tasting the supers should be devised than was adopted at the Palace, it is not every one who would care for honey after another's fingers had been thrust into it.

This has all the appearance of a contrived question by the editor using W. Abbott a relative, as questioner, in order to respond in a manner that would appease Cowan, his friend at that time, and probably following an approach by him. C. N. Abbott was part of the small group that drew up the original draft

Schedule, and even though it had been revised by all members present at the meeting on the 16th May, he would have been embarrassed that it was in large part being blamed for the wrath that Cowan undoubtedly felt, and was making well known. Cowan believed that this incident made him look a fool, not just to the small readership of the *B. B. J.,* but also to those of *The Times.* Cumming was not slow to extract retribution.

The problem was that Cowan had screwed pieces of glass over the honey rendering it not easily accessible, and thus not capable of being tasted. This incident so aggrieved Cowan that I have found five occasions where he writes about it, the last time being fifty years after the event! It is a good example of the unforgiving nature of the individual. At one time he said that surely the judges should have been in possession of a screwdriver, but clearly it was too much trouble for them to remove a screw to taste the honey. It appears that the only instrument they carried was their fore digit! The judges were not named other than that they were some of the vice-presidents. However, Cowan would ensure that it never happened again, appointment of judges in the future and the final say regarding the schedule would be the responsibility of the B. B. K. A., in effect himself.

In class sixteen, *For the best glass super of honey, contents above 20 lbs,* Cowan's entry was placed second of twelve entries; George Fox of Kingsbridge, Devon, was first with the heaviest entry of 50 lbs.

George Fox was listed as a judge, but there is no evidence that he was involved with either of these classes or the disqualification of Cowan's exhibit.

Class 34, for *The best honey extractor,* "....caused no end of excitement." There were four entries, one from Walton, two from Starling who had been selling them for some time, and one from Cowan. Starling was placed first, Walton second. The Walton and Starling machines were very similar, of the tangential type, where the comb is placed at right angles to the radii of the drum in which it rotates. There is further coverage in the chapter covering his equipment.

Despite many claims that Cowan had devised a hive to his own design much earlier, there is no mention that he exhibited it at the Crystal Palace Show. I have also dealt with this matter elsewhere.

The committee of the B. B. K. A. held meetings most months. At the meeting of 8th December 1874, Cowan, who was in the chair, produced a set of rules for the guidance of the committee, and following discussion and modification, these emerge as the more appropriate committee byelaws. The byelaws as passed at this meeting are reproduced in Appendix 2. As can be seen, there is no mention of how the chairman of the committee is to be selected at each meeting, and there is no report of how Cowan became chairman at this meeting. In the etiquette of the day, there was no obvious candidate for the position, and Cowan was probably considered the most senior of the committee men present Hooker, an architect, deputised when necessary. So Cowan was Chairman from the first committee meeting following the first A. G. M., at every meeting that he attended. Whether it was a position sought or obtained by default is unlikely to be known.

The inadequacy of the committee Byelaws is evident and shows the inexperience of Cowan in this type of activity. It is probably the first example of many where he put himself in a situation well outside his 'comfort zone' as the modern parlance would have it. However, it never appeared to concern him, if indeed he recognised it. He often repeated that he considered the B. B. K. A. to be a philanthropic body, and as such, it was not interested in restrictions on it in any form.

The March 1875, issue of the *B. B. J.* contained the requested article mentioned earlier, *Mr. Cowan's System of Working the Bar and Frame Hive,* by the man himself. Of interest here is that he refers to himself as "Mr Cowan" despite having, in the 1871 census, and earlier in the *English Mechanic and World of Science,* claimed to possess a Doctorate. It is unlikely that his peers would have been aware of this. To today's bee-keeper there are no surprises in his method of bee-management, and even in 1875 it was nothing new, similar schemes had been written about by others a decade earlier in the *Journal of Horticulture.*

His description applied to the ten-frame Woodbury hives and an expanded version comprising thirteen frames, even though he later reported that his exhibition supers were Stewerton. The colonies were wintered on seven or eight frames, with 30 lbs of stores, and well ventilated. In Spring he contracted the hive by removing the two outside frames, unsealing the honey in them and allowing this to run down into the inside of the hive, at which point he claimed that the queen will immediately begin to lay. He repeated with other sealed combs of honey in a few days, a process that gave the queen plenty of space to lay in. After this he provided stimulative feeding, expanding the number of frames as the queen increased her egg production. His idea was to stop them storing too much food and thereby restricting the space for brood. When all combs were full of brood, he stopped feeding and put on supers. He used a perforated zinc queen excluder, and split top bars for the foundation strips in the super. In a day or two he added a second super. He provided traps in the supers so that the bees did not have to traverse the brood box on the way out. If the weather turned bad, he removed the supers and fed when he thought necessary. He claimed that in some weeks this gave him 40 lbs surplus. It was, of course very labour intensive, but the gardener no doubt helped.

The above refers to his method of bee-keeping whilst residing at Ryde House, Beckenham, where he was reported keeping bees both inside and out. When the Cowans removed to Hawthorne House, South Street, Horsham, he kept the bees in lofts above his stables:

> Two hives were placed at each window, on tables about three foot high, and the entrances to the hives were about three feet apart, so they were not close together. The only drawbacks to the lofts was the great loss of bees in chilly weather, when many of them returning laden with honey or pollen, dropped down and were unable to rise. It was very easy manipulating the bees, because when the hives were opened most of the bees would fly to the light, and, as the window opened outwards, they could easily be let out and the window closed again. There was very little danger from robbers.

Later in the same contribution surprisingly Cowan states:

> Mr Root works his bees (in bee-house) by lamp-light, but he thought Mr Root could not be serious in recommending this plan. He had tried it, and did not think anyone who had once done so would repeat the experiment. He was very much stung; on opening the hive the bees boiled over, were furious, and flew at the lamp, and most of them did not return to the hive. His first experience of work by lamp was not very encouraging.

But he had already stated that bees flew towards light!

Cowan did not use the quilt, and following his article, there was considerable correspondence extolling the virtues of the quilt, the only support for just using the crown board coming from John Hunter. Yet another example of a topic that has occupied bee-keepers at their winter meetings on many occasions.

The B. B. K. A. committee 'jogged along' it would appear just waiting for the next show, the schedule for which was finalised at the April meeting and published in the *B. B. J.* for May. Cowan was not at either the March or April meetings, but re-appeared for the meeting on 27th May at which the show schedule was revised, the new version appearing in the June edition of the *B. B. J.* This gave editor and printers just three working days to re-jig the magazine to include it, losing a page of other material. The difference in the membership of the two meetings was the addition of Cowan in the third and Abbott for the second. The changes to the schedule are summarised thus: -

- Prizes for all but one class under the heading hives halved, and the certificates were replaced with silver medals. This followed lobbying by the manufacturers who always liked to advertise that a hive was awarded a medal; certificate does not have the same kudos.
- Addition of a new class to account for "special prizes offered by the Hon. and Rev. Bligh and E. Melladew, Esq. for the largest and best harvest of honey in the comb, from one stock of bees, under a system or combination of systems. (Of management presumably). The honey to be exhibited with or upon the hive that produced it (or its facsimile). To be attached thereto, a legibly written explanation of the method adopted, the locality, pasturage, dates of swarming and supering. To this may be added any particulars of his apiary that the exhibiter may be disposed to give, such as number of hives, average yield, etc." The prizes were £5 down, substantially above most other awards.
- Class for heather honey removed – the show was too early to show heather honey of 1875.
- The prize for the best MS lecture on bee-keeping was changed from a silver medal to £5.
- In the class for the best collection of bee-furniture, prizes were changed from money to certificates.

There were some other minor modifications of no consequence. The hand of Cowan can be clearly seen in the changes.

R. Symington, who assisted Abbott in launching the show of 1874 and had been a regular contributor in the early *B. B. J. s*, wrote scathingly about the schedule for the coming show, saying that the committee of the B. B. K. A. had repeated some of the mistakes made at the first show, and suggested some changes, too late of course. More importantly he criticised "some few" who have lost sight of the fact that it was Abbott, and his *B. B. J.* that was responsible for collecting the prize fund for the 1874 show, also all but the final details of the show itself. He continued:

```
Last year many of the judges were competitors, and, strange
to say, the majority of them obtained prizes.  Now there
can be no harm in the judges being exhibitors, but they
ought not to be competitors; the more especially if, as was
the case last year, each has placed in his hands before
commencing his duties a catalogue with the name of every
exhibitor attached to his exhibit.  There ought not to be
the slightest clue given the judges as to the ownership of
exhibits; and were I judge I would disqualify any article
exhibited to which the owner had either attached his name
```

or any other means of identification.

Abbott had referred to Symington's entry as having suffered from an oversight, but he was not the only disgruntled competitor at the show.

The next two meetings of the B.B. K. A. did not manage a quorum, and the next one only just managed it – Cowan this time turning up. There were already letters appearing questioning what the organisation was doing other than run the annual show; a fully justified question.

The second Crystal Palace Bee and Honey Show, took place on 21st, 22nd, and 23rd September and was again fully reported in the *B. B. J.*, with a shorter report in the *Journal of Horticulture*. The poor weather of the previous season had the anticipated result. The bee demonstrations were disastrous due to the poor bee populations in the hives being used, and the honey shown did not match the previous year. The appliances, as would be expected, were not affected, and as always Abbott was not happy with some of the judging in these classes.

Although the honey classes were not as good as the previous year, there were still some very good exhibits and:

```
Undoubtedly the magnificent octagon super from Kingsbridge,
Devon, was the grandest thing of the kind in the show,
being of a gross weight of 93 lbs, and containing 86
lbs net weight of splendid honey in the comb; yet it was
disqualified.
```

It was entered in class eight, the category for the special prizes offered by Bligh and Melladew, and was exhibited fully in accordance with the regulations, but was disqualified despite details of how it was achieved being provided, because:

```
It seems to have been thought impossible that so large a
super could have been filled, according to the stipulations
of the class from so small a hive.
```

As Abbott diplomatically expressed it; stripped of the veneer of politeness, 'Mr. competitor you are a liar and a cheat'. Very serious accusations for the period. He continued, "it will doubtless be a fruitless subject of discussion, and we fear a lasting cause of bitterness"; you can bet on it! At this time it is difficult to judge Abbott's attitude towards Cowan, initially such good friends, the relationship now appears to be cooling:

```
The first prize in this class (five pounds), was taken by T.
W. Cowan, Esq. of Horsham, with a pair of bar supers of a
total weight of 80 ½ lbs., obtained as follows:
```

The stock which produced these two supers is situated in a loft over a stable facing the south in Horsham, Sussex. It was a swarm hived on 3rd May, 1874, and in the autumn of that year a super, weighing 34 lbs. was taken from it. It was then prepared for wintering in its Woodbury frame hive by having two of the outside frames removed from which the honey was extracted, and the combs laid by for future use; two dummy boards were then introduced to contract the size of the hive. The bees were then gently fed until the end of October. The hive was wedged up an eighth of an inch for winter ventilation. A piece of cloth was placed over the feeding hole on crown board. The hive was then left not disturbed until 19th March 1875, when the crown board was removed, and three of the combs were also removed, and the bees brushed back into the hive. The hive was then closed, and on the 11th the five frames remaining were transferred to a clean hive, two of the combs having the cells uncapped to allow the honey to run among the bees. On the 13th the hive was examined, and it was found that the queen had commenced egg-laying. The hive being very strong the remaining combs and their honey cells uncapped, and two frames of comb from which the honey had been extracted placed between the others; the hive was then closed, and on the 25th all the sealed honey cells were again uncapped, and the remaining empty combs placed between the others. A bottle of extracted honey from this hive was the placed on this hive, feeding through three holes. On the 31st March all the honey cells were again uncapped and the honey extracted. And feeding continued until the 20th April, when the hive was again examined. During this time artificial pollen was supplied, which was eagerly carried into the hive until flowers supplied it naturally, when the flour was discarded. On 20th April all the honey was again extracted and the feeding continued, and on the 28th, on examining the hive it was found to contain brood in all the ten frames, and the hive was ready to swarm, queen cells having been raised, these were cut out and a bar super provided, with a bee trap placed on the top of the hive. On 29th the bees were found clustering in super, when a perforated zinc adaptor was placed between super and hive. The super was nearly filled by the 10th of May, when the top was removed and a second super placed over the first. The bees were

then not interfered with until 23rd July, when the top super
was removed, weighing 37 lbs. and a board screwed on the
lower super. On the 24th July the lower super was removed,
weighing 43 ½ lbs and an empty one furnished with guide
sheets put in its place. In this one the bees commenced
comb building, but have not stored any more honey except in
the stock box. This super was removed on 13th September,
and 20 lbs honey has been extracted. Feeding was then
commenced to prepare the hive for wintering, which will
be ready by about the middle of October. Last year my
apiary consisted of twelve stocks, in the autumn they were
all strengthened with bees driven from cottagers hives and
in addition two stocks were made by uniting driven bees
in October and starting them in empty hives by feeding
on syrup. The twelve hives last year produced 707 lbs
of super honey, with 200 lbs of extracted honey. This
spring I started with fourteen hives, three of which had
foul brood. They were deprived of their combs and had to
start afresh. Three others had foul brood and the diseased
combs were excised and the stocks have since done well. By
artificial swarming I have increased my stocks to twenty-
four, sixteen of which are Ligurians and four hybrids. The
honey in this locality is gathered mostly in May and June,
from fruit trees in the orchards and flowers in the meadows.

This full report by Cowan that supported his entry illustrates his method of
bee-keeping at this time. There was, of course, no doubting the honesty of this
entry or any of the other five, but they were all lighter than that of Fox. The
judges did not need to consult their paperwork to know the exhibitor, unless
there was more than one of the entrants who kept his bees in lofts over his
stables in Horsham.

Abbott's comments upon the complications associated with the judges
decision in disqualifying the heaviest super in class eight, that of George Fox,
received support from Thomas Bagshaw in a short letter published in the
November *B. B. J.*, but further comments in a similar vein were rendered unnecessary because at the next B. B. K. A. committee meeting, John Hooker
proposed:

That this meeting deeply regrets the decision of the judges
in disqualifying the magnificent super exhibited by G Fox
Esq., and the position in which that gentleman is placed
thereby. That the committee were not consulted in the
matter, and are of the opinion there was not sufficient ground

for doubting the word and written statement of a gentleman
whose honour and integrity are beyond all suspicion. That
they are desirous of giving to G Fox Esq., an assurance of
their sympathy and esteem, and in acknowledgement of his
great skill as a bee-master, they award the silver medal of
the Society for the super exhibited by him at the Chrystal
Palace Show of 1875.

In the discussion that took place some present thought this gave a 'back handed slap to the judges'. The proposers disclaimed the intention to do so, and the proposition was carried. Fourteen members were present, but not Cowan. There is no evidence to inform whether Cowan stayed away intentionally or for other reasons, but the meeting would have been more interesting if he had been present. Also there is no evidence that Cowan was complicit in any of this, but it is certain that all judges were better acquainted with Cowan than George Fox from Devon, and there was no anonymity. Another exhibitor in class eight wrote asking why he was not awarded a prize but was summarily dismissed. Not a gentleman, presumably.

Of course it did not end there, the judges revealed themselves by writing to the *B. B. J.,* complaining about their treatment. First was R. Symington (he had written earlier about judges being allowed too much latitude), who certainly did not hold back saying that having been:

...asked to be a judge, to pass upon them such a vote of
censure as is embodied in the resolution published by you in
the *British Bee Journal* for this month, shows that although
the Committee as a body may be men of sense and possess
the feelings of gentlemen, there are some amongst them
who are devoid of both qualifications, and who having been
disappointed in the position assigned to their exhibits,
are not able to bear their disappointments in the very
commendable spirit shown by the Hon. and Rev. H. Bligh, (a
reference to the previous year), but must needs vent their spleen by
insulting the Judges.

He continues by pointing out that in the previous year's show he suffered when his "unparalleled exhibit" was passed over and nothing was done about it. Having said that he does not feel at liberty to state why Fox's super was disqualified, but goes on to challenge him as to whether he had not removed the super nine days later than he stated, and that the super had been bred in and contained pollen and brood – but he felt unable to give the reasons for it's disqualification! Exactly the same had happened the year previous, but in reverse, and nothing was done about it, "Mr. Cowan ..... bore his disappointment like a gentleman"; very questionable, but they were determined that there would be no repeat.

Next is Bligh, the junior of the three, who states that his colleagues had the decided opinion that the bees bred in the hive exhibited could not have gathered the honey in the super during the past season. He then adds that he has since received information that proves that they were correct because bees had bred in the super, a fact that had not been stated in the accompanying report.

Third is J. G. Desborough, a month later stating that his decision was based upon calculations made at the time, adding:

> I feel satisfied that had Mr Fox's super been opened at the Show, there would have been no after discussion as to the justice of the award.

There is no evidence that Fox raised the matter, others doing it on his behalf, and the question remains; ultimately, who was first? The schedule was so loosely drawn, and anonymity non-existent, that a debacle was the obvious result. So whether one subscribes to the theory that this was simply a bungled, unmitigated, amateur mess, or to that of conspiracy, history will record a considerable 'fall-out', from this incident.

Cowan's remaining successes were in the classes for equipment. Listed here are the prizes he obtained, descriptions of the various items are given in the chapter on his bee-keeping inventions.

Class 24. Best bee feeder. Cowan had an entry in this class, but was not placed. Reference to a description of the device will explain why.

Class 25. Bee quieters. Cowan first of just three entries. All adaptations of every-day items.

Class 27. Supers for producing honeycomb in saleable form. "...the ingenious Mr Cowan being the distinguished exhibiter". Cowan first.

Class 28. Honey extractor. Cowan had three entries, 'Cottagers', 'Amateur's', and Rapid'. Cowan first with 'Rapid', even though Abbott considered it deficient in some aspects.

Cowan made a contribution of £5 to the prize fund and won at least £9, but the kudos was more important, it was never about just taking part for him.

Class eight was not alone in attracting criticism of the judging. Class 32 for the best MS on bee-keeping, with a single prize of £5 attracted just four entries, Desborough, W. Hunt, J. N. Coleman, and Rev. J. D. Glennie; Hunt being declared the winner. At the B. B. K. A. meeting following the show, Desborough asked when it would be published. It transpired that only Hunter and the judge had read the winning entry. The secretary was called upon to read the MS, but part way through, Cheshire put a motion that it was not "of sufficient merit" to publish. This was passed, and the "judges decision was rendered nugatory". Desborough was to pursue this in the future, but without succeeding in producing a satisfactory outcome as he saw it; he was now in the wrong faction. He had better fortune in another class as Abbott reported:

```
In class 19, liqueur or wine, there were five entries.  Won
by Desborough with two bottles of sack mead, the reversion
of one bottle he left for us: but only by the odour of
the bottle could we possibly judge of it's quality as some
'practical joker' - who ought to have been detected - had
emptied it.
```

Of two bottles of a 'honey drink' again given by the exhibitor to Abbott one had been drunk. Alcohol, was not the only thing stolen, the "highly ingenious multiplying gearing and handle of Novice's extractor" (from America) disappeared, as did Cowan's MS book with red covers, containing his apiarian notes for some years past. Now, who would have done that? In his autobiography Root informs us that his gearing was from an apple parer.

The show was the culmination of the years work for the B. B. K. A. committee. Given their inexperience and the many personal rivalries and egos involved, it was a considerable achievement. However, there were times when it appears to have descended into a total shambles, and certainly lacked credibility, most of the prizes being awarded by members of the committee to other members. The conclusion of the second Crystal Palace Show marked the end of the first year proper of the B. B. K. A. The A. G. M. was chaired by Cowan, but the report in the *B. B. J.* was very short because the paper work had been "lost on the railway". Hunter resigned as secretary.

Of the six reported B. B. K. A. committee meetings between October 1875 and March 1876 Cowan was only present at the February meeting and little business was transacted, reinforcing the opinion of many that the Association was only interested in organising an annual show. When the time came to devise the schedule for the next show, Cowan re-appeared and attended on a regular basis. The show for 1876 was to take place at Alexandra Palace, the schedule being very similar to the previous year, the changes again, clearly indicating the influence of Cowan. The resignation of the secretary appears to have caused considerable disruption, and it was only at the August meeting, three weeks prior to the exhibition, that Judges were chosen and a resolution passed that no individual could judge a class in which he was an exhibitor; obviously they had done at the two previous shows. At what seems to have been a meeting called in haste, it was further resolved that no exhibit be disqualified by judges without reference to the committee; so none of their exhibits will be disqualified then! Many of the arrangements for the show were haphazard, last minute, and left to the hosting venue, illustrating further the poor organising skills or lack of energy of those involved. Enthusiasm for the Association and doing the necessary work for the show was beginning to wane.

In view of the poor organisation it is no surprise to learn that the display tables were erected late, not something any member of the committee would do or probably know how to do, judging was therefore late, catalogues were late from the printers, and the awards were not available until mid afternoon on the second day. The bees purchased on behalf of the committee were from hives with foul brood and were very poor, making the demonstrations difficult, many members who had not agreed to help in this area assisted including Cowan. Few lessons seemed to have been learnt from the first two shows, and the Association lacked someone with real drive.

Abbott reported that, "judging considered accidental" and "did not give universal satisfaction". It would be surprising if it were any different, judges were chosen for their position rather than ability. My subject is mentioned as follows: -

Class 4 – collateral hive, Cowan placed second, and bronze medal, with a hive on the Giotto principle. A genuine Giotto hive was unplaced.

Class 11 – largest and best harvest of honey in the comb from one stock.... Same as last year' class 8. Cowan was again first with a weight of 120 lbs, even though an entry from P. H. Phillips was heavier at 131 ½ lbs, it was placed second – beaten on quality, third prize did not have a weight appended. An entry of three cwt. made by Captain Martin, did not materialise "through some defect in the railway arrangement." Another entry of "great weight" was delayed by the railway and arrived too late. Would probably not have mattered.

Class 12 – best exhibition of super honey from one apiary. Cowan first of six entries, with seven boxes and a vase, weight about 150 lbs. Second, Phillips again, with sixteen supers, weight about 250 lbs. "...causing considerable surprise".

Class 14 – best super of honey, twenty-five entries. First Walton, a cottager, with 75 lbs, Cowan fourth! A few years later Walton was visited and questioned by Peel to confirm that he was a genuine cottager, because it was thought impossible for a cottager to be so successful with bees, especially beating the Chairman. Following the inquisition, Peel/B. B. K. A. issued the financial criteria that would define a cottager in future.

Class 17 – for the best exhibit of honey in supers or sections of supers, separable, and each of not more than 3 lbs in weight, the total weight of each entry not to be less that 12 lbs. This was a new class and one cannot help but feel that Cowan put this in having developed the equipment for doing it. Cowan first and second of just five entries.

Class 29 – best honey extractor, portability, and cost to be taken into consideration. Thirteen exhibits, first and silver medal, Cowan with 'the rapid', which had taken the honours at the previous show. It was the highest priced at £3 and "certainly not the most portable machine". Abbott believed the silver medal should have been withheld because there had been no significant changes to the winning entry since the last exhibition. Cowan also exhibited his 'Express extractor', price 55/- which was highly commended

Class 30 – beeswax Cowan second of nine exhibits.

Detailed descriptions of all Cowan's equipment entries are included elsewhere.

Cowan subscribed £5 to the prize fund, received £11, one silver and one bronze medal. There was little criticism of the judging in the pages of the *B. B. J.*, an indication that most people were satisfied or not prepared to waste time on a cause they believed to be futile. Charles Tite wrote objecting to the criticism of the judging by Abbott, saying that the incorrectly awarded medals would be declined, because they were awarded to men of honour. None were.

As if mimicking their bees behaviour, the B. B. K. A., became relatively inactive in the winter, yet again, but it was becoming apparent that there were insufficient funds to run a show in London during 1877. Membership levels were falling, the starting of local associations having an impact as well as dissatisfaction with the central body, and there was a reluctance of many to once again subscribe to the show fund. In the final list of subscribers even Cowan's name was missing, although he had contributed to the fund for a tent to hold demonstrations of bee handling.

A group of members of the committee had become unhappy with the manner in which Abbott was increasingly using the *B. B. J.* as a vehicle for his bee business, although he always inserted contributions from his rivals, describing their equipment, and did not charge for publishing anything the B. B. K. A. sent in. These members wanted the B. B. K. A. to have it's own magazine "untrammelled by commercial interests". Not surprisingly there was a major "falling out". Abbott withdrew his magazine as the official organ of the Association which responded by seeking subscriptions for the establishment of its own magazine.

C. N. Abbott submitted his resignation from the association that he was primarily responsible for establishing, at the meeting on March 9th, 1877. Letters praising C. N. Abbott and criticising the B. B. K. A. then started to appear in both the *B. B. J.* and *The Journal of Horticulture*.

The B. B. K. A. was in terminal decline. There was drastic contraction of membership, inadequate funds, lack of energy, direction and leadership. It was reduced to awarding medals for regional shows because of its inability to mount a show of its own. At what appears to have been a crisis committee meeting in July 1877, Cowan although present, was not Chairman, the chair being taken by E. Walker, the prime motivator of the proposed new magazine. If there is any significance in this, it is unclear. Most of the meeting was occupied with the secretary reading a report on the state of the Association, that he had been asked to prepare at the June meeting. It was not pleasant for those few members present.

As a further indication of the increasing crisis within the B. B. K. A. the A. G. M. was not held as due in September but on 13th December of that year. There were 10 members present only one of whom was a non-committee man. Even though the meeting was three months late only an 'approximate balance sheet' could be presented. The income was £89 19s 9d, including donations to a prize fund for a show that was not held, and a tent that had not been bought, expenditure £76 15s 10d; a sorry state. But how could they spend such a large sum of money? £67 10s had been promised in response to the appeal for funds to launch their own magazine. The rest of the report was as per the previously referred to July meeting. The meeting was adjourned. The report content is very well expressed by Abbott's footnote, having clarified the financial situation he writes:

> All that has been done has been the awarding of several silver and bronze medals; a correspondence with the conductors of the Paris exhibition; and an unsatisfactory correspondence in regard to the International Show at Carlisle, and although it is said nothing decided had resulted from it, a great Bee and Honey show actually took place, and thousands of persons were made acquainted with bee manipulation through the *enterprise* of private individuals. The remainder of the report is made up of regret, except that alluding to the attempt to establish a journal "which should be untrammelled by trade interests", the editor of which is to be 'able and responsible' and the capital raised by general subscriptions. Is there some huge '*private interest*' in

the background that hopes to obtain a public position with
public money; and so achieve a new eminence to hail from.

This has very similar tones to Abbott's writings to the *Middlesex Times* prior to his starting the *B. B. J.* It appears to convey the opinion that Abbott was not 'able or responsible'.

In the editorial for the same number of the *B. B. J.*, Abbott claimed that before he received the B. B. K. A., A. G. M. report, he wrote of the rise in membership when he was involved, and the fall since he left the Association, stating that it was in "a moribund condition and must shortly be 'wound up'". He was not happy that bee-keeping should not have a central body and proposed forming a new Central Association of Bee-keepers, stating it's objectives and suggests that Grantham is the best place for it to be situated. The objectives for the new society are given and he calls for promises of support. So, there is one faction with a magazine but no association, and the other has the Association (just), but no means to start its own publication. It is difficult to detect whether Abbott was serious about his proposal or if it was posturing. It does not seem likely that he would wish to go through the exercise again, but several letters expressing support were published.

The B. B. K. A. suffered a further blow in January 1878 when Fox Kenworthy the secretary resigned. It should not have been a surprise because he was a relative of Abbott and only 24 years of age at the time. Hunter stood in yet again, and produced the Balance Sheet of the Association which was circulated mid March and included notice of the adjourned A. G. M. It indicated that the present Committee members wished to stand down and have a new executive appointed, and if this does not happen the Association would be wound up. Abbott published the Balance Sheet, pointing out the absurdity of continuing a society where £28 15s 6d had been spent in expenses to manage £26 1s. At the meeting seven committee members attended and three Association members, one of whom - Rev. H. R. Peel, was Chairman. Although very few members expressed the wish that the Association continue, that was the decision of the meeting, with all the committee present continuing; Peel becoming secretary and agreeing to meet with Abbott to solicit the use once again of his Journal as the official organ of the Association. Peel persuaded Baroness Burdett-Coutts to become President. Dissident members had been purged from the committee to Abbott's satisfaction. There was to be a show in London in 1878, and the rules regarding membership revised. Harmony had broken out!

Comparing the two lists of members before and after Peel's appearance in order to identify the dissidents is not completely conclusive, but would indicate they included Rev Hill, Clark, Edwards, Neighbour, and Walker.

In all this mayhem it is difficult to determine the part, if any, played by Cowan, but he later revealed that he was one of those in favour of the Association continuing. Although he chaired many of the committee meetings prior to the troubles, there were also occasions when he was present but not voted to the chair. However, it is obvious that he was not a sufficiently strong and powerful individual to lead such an organisation in troubled times. Later he would admit to having difficulty dealing with the warring factions within the Committee at this time. Ideal Chairman material then. However, he would flourish behind Burdett-Coutts and Peel, two class heavy weights.

The year of 1878/9 appears to have been one of improvement in the organisation of the B. B. K. A. Peel injected considerable energy and enthusiasm. The central show for the year was held at South Kensington in early July. Cowan was not a judge and did not feature in the awards, although it is not clear if he was an exhibitor. He was at the B. B. K. A. meeting on the first day of the show and was voted to the chair. He had only contributed £1 1s to the prize fund. Abbott refrained from expressing any opinion on the judging, and there were no letters published criticising any aspect of the show, or none that were published.

There is little doubt that Peel had generated a greater sense of purpose in the committee of the B. B. K. A., especially Cowan. In addition to the show at South Kensington the Association was also to be involved at the Royal Agriculture Society show at Kilburn, with an exhibition only a little smaller than the main one. Quarterly meetings involving representatives from the Counties and including a Conversazione, were now a regular occurrence, and committee meetings transacted some real business.

The members of the committee were now being elected, but the method of allocating the number of votes that each member could cast, led to the criticism that the committee elected itself, borne out by the results. At the April 1879 meeting it was decided that the Chairman of the committee would be the person who polled the largest number of votes. Abbott was the only one to vote against this resolution.

At the same meeting there was a conversazione at which Cheshire presented *Abdominal distension of the hive bee during Winter and the means of checking same.* Many individuals contributed to the discussion, mostly with short statements or questions but Cowan had come well prepared, and described in detail his method of wintering bees. The following is Abbott's version of the Cowan's contribution:

> He agreed with Mr Cheshire as to the production of heat by
> the consumption of honey, and likewise as to the nitrogenous

food, and the thing was to devise a method of keeping bees as nearly as possible at a certain temperature. The first point was to have the bees strong in Autumn. Many hive could not pass through the winter because of the limited number of bees. The next point was to consider the best means of ventilation. Certain gasses were given off, and unless the hives were well ventilated the bees consumed these gasses. This year he had tried a plan of his own, the idea being suggested by the Gordon fire grate. The fresh air was drawn in, and passing through a warming chamber was brought to a certain temperature, and it was reckoned that the whole of the air in the apartment was changed in from 30 to 60 minutes. In a hive the quantity of carbonic acid gas given off from consumption of sugar or honey was such that if an ounce of honey were consumed in a day, the hive would require to be ventilated every quarter of an hour. As a matter of fact, bees did not consume an ounce of honey a day. When the outside temperature was forty the temperature of the cluster of bees should be sixty-five, and it might be supposed that the hive would require to be ventilated every hour. To maintain that temperature he contracted the sides of the hives to three four of five frames. To get rid of the gasses he had a little tin box about an inch square, made, from which a tube passes to the top of the hive and branched off to the two extreme ends of the tube being perforated. At he bottom of the square tin box he had another tube which passed out at the bottom of the hive and affixed combs on each side, compelling the bees to form the cluster round the box. By that means the temperature was kept up to within 10 degrees of the heat of a cluster throughout the winter. From 55 to 65 degrees had been the temperature of his hives through the winter. To get rid of the gasses he had a funnel to within an inch of the top of the hive; the top of the hive was closed, but free air came in with the foul air, which escaped through the openings around the hive. The lighter gasses would pass up the funnel by the heat caused by the cluster round the funnel. In that way he had been able to keep up the temperature of the only hive in which he had tried the plan, and this hive was one of the strongest he had. Nearly all the combs had brood in them which was a proof that ventilation and warm-air ventilation was necessary.

In a follow-up to this Cowan said that he approved of double walled hives and Peel contributed a bit of nonsense, conveying the information that his hive was kept in his study, therefore the bees were at a constant temperature, and that he used a Cheshire double walled hive made by a local carpenter. But he was not a bee-keeper in the usually understood sense.

With W. Carr, Cowan was appointed judge at the 1879 Kilburn show. I believe this to be his first such appointment. At this show there was an exhibition of two cwts. of honey from Thurber & Co of America, that was awarded the highest prize, and their agent, Hoge, claimed that he had received an order from H. M. the Queen. This was an incident that caused considerable irritation to Cowan over the ensuing years and will be returned to. Royalty visited the show, but more importantly so did T. G. Newman, editor of the *American Bee Journal* and C. T. H. Gravenhorst, editor of *Illustriere Bienenzeitang* and inventor of a hive that bears his name – a straw structure with moveable frames, that had to be inverted to remove the frames. These two individuals, with others, were entertained by Cowan at Comptons Lea at a substantial gathering of beekeepers. In the *A. B. J.* Newman reported that he was entertained at the palatial residence of Peel and attended a meeting at the "fine residence" of Cowan. A few years later it had been downgraded to "Mr. Cowan's country cottage"! The report submitted to his Association in America was not very complimentary to British bee-keeping, including:

> America stands first in the World for honey production as well as for scientific management and improved implements for the apiary.

He submitted it to the *B. B. J.* "hoping it met with approval". How wrong could he be!

At the October 1879 meeting, Cowan, at the request of Peel, gave his first paper at a B. B. K. A. conversazione – *Wintering Bees*. It followed his contribution to Cheshire's paper at the previous such meeting, and was a more practically based offering, but an undoubted snub to Cheshire. He stated that the requirements for successful wintering were: - "A uniform temperature; a sufficient quantity of food; a large number of young bees; and a hive free from moisture and noxious gases." To address these points he said that a minimum of 25 to 30 lbs of stores was necessary; cut passageways through combs; induce late breeding by stimulation feeding and providing room for it; using Cheshire's divisional crown board for ventilation; having a quilt that not only covered the top of the hive but also extended down the sides, enabling the better dispersal of moisture by capillary action. On another occasion he said that he did not use a quilt, just the cover board. He described details of two separate experiments carried out on single hives, drawing such scientific conclusions as "Both wintered successfully and turned out strong in the spring". He then described the extraordinary device for changing the atmosphere in a hive without reducing the temperature and this is detailed in the chapter on equipment.

At the November 1879 committee meeting it was decided to purchase the diagrams that Cheshire exhibited at the Autumn show at the price given in the catalogue, reproduced to a size 40" x 27" using a four colour process and sell for 5/-. There appears to have been no debate as to copyright ownership, or whether Cheshire was to be credited. These diagrams and their ownership will recur, with Cowan much involved. The finished product was shown in February of the next year.

Although the membership of the B. B. K. A. (158), was still less than half that at its peak, the committee in general and Cowan in particular, were much more active. They had met with the authorities of the Royal Horticulture Society to organise the next show; the preparation of the handbook for cottagers was half finished - Cheshire still to do his part; it was decided to "spread into Ireland" by sending two experts and the bee tent for a tour of the island in the Summer of 1880, to teach Irish bee-keepers the new methods, and Huckle, the assistant secretary would now be paid. Cowan returned the highest number of votes in the committee elections, and was to be Chairman for the forthcoming year. At the conversazione he expressed his preference for the Ligurian bee. He was now a very regular attendee at the various B. B. K. A. meetings and events.

As with the previous year 1880 saw Cowan having ever more influence and say in all matters bee-keeping. At the conversazione which featured Hunter's paper on the future of bee-keeping and the best way to educate cottagers, Cowan was active in the discussion, expressing his condemnation of 'American' hives - single walled and only cheaper because timber was less expensive in that country. Throughout this narrative there are many examples of his fierce Nationalistic character, about the only exception being foreign equivalents of himself, mostly whom he considered friends. This was not abnormal at the time – he was a national of Great Britain and it's Empire and proud of it. During the same conversazione, he said, of how to introduce cottagers modern methods of bee-keeping, "To give a cottager a bar-frame hive and tell him to stock it with bees was about the most harmful thing they could do", and that they needed to be taught the basics of bee-keeping first. This is in direct contradiction to his earlier pronouncements on the matter and, as is usual, there is no acknowledgement of the change or what brought it about. There will be several more such examples in this work; a continuing theme.

Cowan was now contributing less to every appeal, normally £1 1s, down from his usual £5. This is more likely to be the result of Burdett-Coutts and Peel joining the Association and the necessity for him to contribute less than them, rather than financial restraint. The financial loss to the B. B. K. A. of the Irish expedition was £45, made good by Burdett-Coutts (£25) and Peel (£20). Much later Cowan claimed that he also contributed.

Cowan exhibited at the South Kensington show, (1880), and in the honey classes only managed a third for his sections, but exhibiting was anonymous now. He was first of six entries with his easily recognised extractor, and only commended for his swarm box. Net return was a silver medal, 15/- and a certificate. A cynic might observe that his contribution had reduced in proportion to his winnings! In their report on the decision in the class for extractors, the judges said that there was no hesitation in awarding the silver medal to Cowan, he having improved the machine shown last year in respect of the mechanism for reversing the combs, describing in detail the mechanism. Most of the other exhibits had not been changed from the previous year. It is interesting to note that the judges for the extractor class were members of the B. B. K. A. committee, and those for the honey classes were not.

At the third conversazione of 1880, Rev. E. Bartrum, Head-Master of Great Berkhamsted School, an establishment where later Cowan would place his eldest Son, read a paper on the Stewerton hive. It was an octagonal hive invented by Robert Kerr in 1819, and derived it's name from Kerr's home town of Stewerton, North Ayrshire, and had been much lauded in the writings of 'The Renfrewshire Beekeeper' and others from Scotland, quoting very large crops of honey when using it. Cowan was the first to contribute to the discussion, it was lengthy based upon the experience he had had over 16 years of using the Stewerton hive. He omitted to say that he had only one such hive during that period. He did not totally condemn the hive, but said that he considered it to hold an intermediate position between the skep and a bar-frame hive, whatever that meant. He continued, giving its shortcomings as he saw them. The fact that an Englishman was asked to present the paper was not greeted well North of the Border, as would have been expected, and in a long letter 'The Renfrewshire Beekeeper' after correcting certain statements made by Bartrum, took Cowan to task on the contents of much of his contribution, other Scottish Bee-keepers joining in later. This was the first time that Cowan's outpourings on bee-keeping management had been criticised, and the quarrel was to last until death intervened. However there could be no doubt on the English side of the border at least, the Stewerton hive had 'had its day'. The shape meant that it was difficult to interchange frames, and despite claims to the contrary, the arrangement of sliding bars between frame tops, must have been subject to propolising, and effectively it had already been superseded by the Carr-Stewerton hive. John Keys in 1780 and Thomas Wildman, 1767 had already used sliding bars, and the shape had been around for even longer. It was not really anything new.

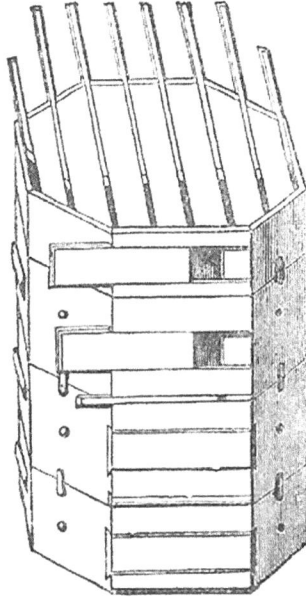

The Stewerton hive from *The Honey-bee*, W. H. Harris

Modern bee-keeping, a handbook for cottagers was eventually published during the year, and quickly sold out the initial print run of 5,000. It is interesting to note that the story behind the engraving for the frontispiece was changed from it's initial appearance in the *B. B. J.* With one exception, Cowan attended all the B. B. K. A. meetings and shows in 1880. Peel had asked to be relieved from the position of Secretary, but agreed to continue if the Association wished it, and Huckle was prepared to continue also. The reason cited by Peel for wanting out, was continuing poor health, a pointer to future events. However, it had mostly been a good year for the Association and the Chairman of its committee.

The B. B. K. A. continued to prosper throughout 1881, increasing membership, monthly committee meetings, quarterly conversaziones, London show, greater number of affiliated County Associations, publication of Cheshire's drawings, and the second edition of *Modern bee-keeping*.

MODERN BEE-KEEPING:

A HANDBOOK FOR COTTAGERS.

PUBLISHED BY
THE BRITISH BEE-KEEPERS' ASSOCIATION.
1880.

HIVING UNDER DIFFICULTIES.

Frontispiece and simple title page of B. B. K. A. *Modern Bee-keepeing*

The B. B. K. A. continued to prosper throughout 1881, increasing membership, monthly committee meetings, quarterly conversaziones, London show, greater number of affiliated County Associations, publication of Cheshire's drawings, and the second edition of *Modern bee-keeping*.

Cowan's influence was increasing all the time, ever present at B. B. K. A. events, he was at the centre of most decisions. At every conversazione he dominated the discussion no matter the topic, continually pushing his views, often disagreeing substantially with the speaker and other contributors to the discussion. These offerings provide one of the best insights into Cowan's character and sometimes include valuable information not to be found elsewhere. 1880 and 1881 were defining years for my subject, putting him at the head of British bee-keeping.

In all organisations when a strong and popular leader emerges from the masses there are causalities. Schoolmaster Frank Cheshire who had already published *Practical bee-keeping* in 1873, enlarging and revising it in 1876 and had allowed the Association to purchase 'his' diagrams and publish them, was the first to be 'put in his place'. He had already had public disagreements with C. N. Abbott and Cowan on matters of bee-keeping principles, but it was recognised that he was doing valuable microscopic work, and was probably the best scientist on the committee. The *Journal of Horticulture* had reported "the author of *Modern bee-keeping* is Mr Cheshire, who wrote it gratuitously for the British Beekeepers Association". It was not known if Cheshire, who was sub-editor for the bee content in that magazine, had written the entry. Because Hunter and Cheshire had both written books previous to this they were chosen to write the draft of *Modern bee-keeping*, which was then finalised by a small sub-committee including Cowan. Critically Cheshire had drawn some of the illustrations for *Modern bee-keeping;* the B. B. K. A. then having them engraved. Cheshire wished to use some of the engraved blocks for use in his new book and claimed that the copyright was his, also that he should be allowed to claim authorship of *Modern bee-keeping* because he wrote all of it. Amazingly, the committee led strongly by Peel, considered the copyright belonged to the Association because they had paid for the engravings, and that because all the committee had seen and made suggestions for the improvement of the book Cheshire could not claim authorship, and as Cowan put it, "if they (the sub committee), did not write the work they gave their brains to it....". Later Abbott would claim that the many improvements that he suggested were ignored.

Regarding the blocks it was suggested that Cheshire make application to the Association to use them; apparently Cowan had already done this with one block. There can be few better examples of the arrogance of the committee led by an aristocratic autocrat. It was a matter that would return in a different context twenty-five years later and is covered in a subsequent chapter.

Cheshire resigned his position as Librarian and gave back the Cyprian queen that had been placed with him because he had failed to breed pure Cyprian queens from her. Exactly how he was to achieve this went unexplained.

However, as might be anticipated the 1881 annual show at South Kensington did not pass without controversy and problems. In the class for 12 x 1 lb sections, Cowan was placed first, prize £1. However a "slight of hand" as Abbot described it, was applied to Hooker's exhibit. He had staged twelve perfect sections but upon judging, one had been replaced by a very inferior effort, the reverse switch taking place after the judging – "a shameful piece of rascality". The Association had implemented very strict regulations governing who was allowed to be in the room at the same time as the judging was taking place. Forward the usual suspects then! Cowan was third (5/-), in the class for 12 x 2 lb sections, and in the extractor class he took first (silver) with his 'improved automatic' and second (bronze), with his 'Rapid'. Two new classes had been introduced, (i). Transportation crate for sections, (ii). Transportation crate for jars of honey; Cowan was place first in both out of four and five entries, bronze medal in each case. Hooker's twelve sections were placed first at the Dairy Show later in the year, at which Cowan was a steward and did not compete; honours equal.

In the beginning of what was to be a long running problem for Cowan, the application to allow the sending of Queen's by post was rejected by the Postmaster General. The committee continued to tinker with the rules and regulations of the Association but also continued to ignore them if it did not suit.

In addition to his show activities Cowan was, like many others experimenting with the various races of bees that were being imported at this time. In 1875 he purchased several Carniolan queens from Mr. J. Hunter who imported them. He found them to be swarmers, "…. they would even swarm with supers on, but by giving them plenty of room we were able to control them." He asked for reports from others who have this race of bees, but he was only seeking agreement, not a genuine discussion of their merits. He first encountered Cyprians at the 1875 exhibition, when he received a "real stinging" from a colony on show by Neighbour. This did not deter him and from "a hive or two" in 1875, by 1882 half his apiary were Cyprians and he stated that for two years he had twelve pure Cyprian stocks from imported queens sent by reliable dealers:

> ………and should have had no difficulty in keeping the race pure to this day had we wished so ……… any one taking the trouble can secure pure fertilisation if he wishes.

He concluded that their variable temperament meant that they were only suitable for the expert beekeeper, pointing out that Raynor, the real authority on them in Britain, was of the same opinion. Even with careful breeding he had not been able to breed out the sudden vindictiveness they exhibit. In his usual frustrating manner the reader is supplied with the problem, informed that he has studied it and experimented upon it, and then gives the conclusions. No detail is provided as to the process gone through to arrive at those opinions, even the basics such as number of colonies studied, number of years of the study, are not given, not to mention the variable parts; season, colony strength etc. For today's beekeeper it would be very educational and interesting to know just how he did his research so the thoroughness of it could be assessed in conjunction with the techniques of his day. Most beekeepers would be keen to know how he could keep the race of a honey-bee pure in 1880, and how he assessed that purity. Such behaviour still exists today among bee-keepers; claims made for devices and bees after minimal research; the rush to publish the results in order to lay claim to that finding uppermost in the mind of most, despite the limited value of such information or it's validity. He wrote two short chapters on these bees in the *Guide Book* but again it was just a summary.

By the end of 1881 Thomas William Cowan was firmly established at the head of organised beekeeping in Britain. Since the establishment of the *B. B. J.* in 1873 and the formation of the B. B. K. A. a year later, his influence on all matters beekeeping had increased year on year. However, during his stewardship the Association effectively collapsed, and even though he was one of the twenty members that voted to continue, a position probably influenced by Peel joining, he had not been able to stop the decline. Initially, it was Abbott who provided the impetus and the second time it was Peel. Cowan's effort was much greater after Peel became involved as secretary and persuaded Baroness Burdett-Coutts to be President. The presence of the upper echelons of society at the top of the Association, of which Cowan was now firmly established as Chairman of the Committee, and the ongoing formation of County Associations to reach down to cottagers, positioned Cowan in his comfort zone.

Chapter V

An Established Bee-keeping Figure

At the beginning of 1882, Cowan was forty-two years of age, had lived in Russia, completed an engineering apprenticeship with a major company; taken over Kent Iron Works from his father and lost it; had his name attached to several patents; exhibited at Alexander Palace show of 1862; married the daughter of a wealthy Sussex businessman; had five children; built a mansion at Comptons Lea, Horsham, Sussex; bought Tryfynant mine, and failed with it: published two editions of *The British Bee-keepers Guide Book;* become Chairman of the committee of the B. B. K. A. and a respected authority on bee-keeping. Also, if he is to be believed, he had been consultant engineer on several major projects, obtained a Ph. D. and M. A. Most importantly he and his wife had inherited enough money to give them a comfortable life without recourse to work.

In the 1870s, the topic of a standard frame and hive was the subject of much discussion in the *B. B. J.* but the B. B. K. A. appeared reluctant to get involved, until, at the A. G. M. for 1882, the following resolution was proposed:

> That it is desirable that the B. B. K. A. do set forth a *Standard Frame* stamped by its sanction and authority, with the view to bringing such frame into general use. It's size and form to be determined by a Committee appointed for that purpose.

This appears to indicate that the full Committee were not to have the final say. Rev. G. Raynor, at the first conversazione of the year, presented a paper on *Bee-houses and Hives,* in which he pre-empted the thought process that this Committee should follow, concluding "....that there is a general consensus of agreement in favour of a frame nearly approaching, at least, the Woodbury in size". Cowan, as usual contributed at length, stating that he:

> ......remembered when the question of a standard frame had been discussed by the Committee: and it was abandoned because the hive-makers who had been consulted could not agree with the Committee as to size. The general feeling seemed to be in favour of the Woodbury size, which was 13" x 7 ¼" inside measure and he himself should prefer a frame as nearly that size as possible. In Sussex he had found the best results were obtained from the Woodbury size frame.

Very much later in his life he said that prior to the decision on the standard frame, that he had experimented with several different size frames before coming to that conclusion. Evidence from his own pen, shows that he had little other than Woodbury size hives from a very early stage. He also came to the conclusion that the Stewerton hive was unsuitable after using just one for a few years. He exhibited a modified Giotto hive at an early exhibition, only to declare later that he "….was obliged to give it up in disgust, as it always got propolised and the frames stuck together". Despite the impression Cowan gave, his experimentation was shallow.

At the A. G. M. February 1882, the sub Committee to determine the size and style of the standard frame was chosen. It comprised C. N. Abbott, Cowan, F. Cheshire, J.G. Desborough, J. M. Hooker, Rev G. Raynor, A. Neighbour and Rev F. T. Scott. Abbott and Neighbour were suppliers of hives; all the others had devised hives and published details of the frame size. Only Neighbour was using a frame substantially different from all the others, making the result predictable. None of the participants operated a large number of hives. Only two meetings, chaired by Cowan, were required. Decision allegedly unanimous, was 14" x 8 ½" outside measurement, width, 7/8", top bar, 3/8" thick, side bar ¼ " thick, bottom bar 1/8" thick. The dimensions were announced at the South Kensington Show in August, seven years after it was first proposed. There was, as could be anticipated, much disagreement, mostly surrounding the belief by many, that the internal size should have been such that it would accommodate six sections, not that they could agree as to the size and shape of the section to make that a feasible criterion, although the American 4 ¼" square version was already beginning to dominate. The reason for this desire was that it was practice amongst many beekeepers to start sections on the outside of the brood nest and then move them above it for completion.

The matter of a standard hive was also discussed but it appears Cowan was not in favour because it would have eliminated his thirteen-frame hive that he was using at the time. Abbott was not happy that the sub-committee did not specify the distance between frames and method of spacing that was his passion at the time, Cowan expressing the opinion that how frames were spaced was irrelevant. On another occasion he had said he spaced his frames by finger and eye, and he wanted flexibility.

Despite the opposition at the time and continuing discussion amongst bee-keepers every time they now meet, the chosen size has endured for over 120 years although others are now commonly available and considered alternative standards.

Mid 1882 Cowan and B. B. K. A. were forced by "Platelayer", a cottager's pseudonym, to concede and change the classification and examination of experts, creating 1st, 2nd and 3rd categories. Cowan was not happy with the change, but the weight of opinion coming into the *B. B. J.* left him with little choice, not that he always fell in with public opinion.

1882 marked the last time that Cowan exhibited, or more accurately was recorded as a prize-winner at exhibitions. In the early part of the year he exhibited at the Cardiff show, the only recorded time outside London. In the class for the best 12 x 1 lb jars of honey he was placed first and second. The judges were Peel and Hooker. This was the first show that he exhibited and was placed in a class for hives. In the class for the best frame hive of a substantial nature, max price 15/-, he was placed third, and for the best frame hive for cottagers, max price 10/-, he was also third. Judges were Filleul, Bevan Fox, Tite. In all B. B. K. A. supported shows, the hives exhibited were supposed to be available for purchase at or below the price limit of the class. In Cowan's case how this could be is difficult, because he did not manufacture and sell any of his inventions. The first of the hives mentioned was his double-walled hive, described in every edition of his *Guide Book,* but there was no description of his cottagers' hive here, or elsewhere that I can find. The double-walled hive was manufactured and sold by others with due acknowledgement, but never for a price of 15/- or below.

At the South Kensington show, late Summer 1882, Cowan was second with his rack for 1 lb or 2 lb sections; first and second (bronze medal and certificate), for mead accompanied by recipe; first, second and third (silver and bronze medal) for his extractors; first for honey jar; first second and third (silver and bronze) for his transportation crates; and first (silver medal), for the best invention - a machine for fixing foundation into sections.

The criticism of the judging at Cardiff and South Kensington shows came from many correspondents to the *B. B. J.,* particularly the editor. Essentially they were no different than had been levelled before, but this time they were bitter and directed in particular towards one individual – Cowan. His two hives could not be produced for the price quoted, a point that Abbott had long campaigned about. In the class for mead his awards were for mechanically aerated water, sweetened with honey and flavoured with fruit juices. Since when has this been classified as mead? These drinks had been a quest for Cowan, a teetotaller, for some time as mentioned elsewhere. His extractors had no improvement since they were exhibited the previous year, and therefore were not entitled to the medals. His winning honey jar was "a horrible heavy, green-glass jar"; the apparent justification for the award according to one of the judges was "you can use them as a hammer to drive a nail in with". His transportation crates were as exhibited the previous year, and therefore not entitled to the medals under the rules, and there were others shown less than half the price, something that was supposed to be taken into account. "He also won with an American foundation fixer of his own invention", and that had also been exhibited before. Previous criticisms of the judging had been veiled, but this time there was no doubting where the attack was focused. One particular letter read "had the exhibitor been some other person (than Cowan), one prize would probably been held sufficient acknowledgement of his genius". He continued criticising the impartiality and qualifications of the judges and the anonymity of the exhibits, and suggested that the judges should undergo an examination in the same way that experts do. He gave his name and address, but presumably he entertained no ambition as an exhibitor or candidate for the Association Committee!

There was little doubt that Abbott was once again becoming disillusioned with the behaviour of the Committee of the B. B. K. A. and this could have fuelled his unusually bitter criticisms. He was not alone, and even though it could be expected that a fledging organisation would not get something like the regulations for, and judging of, large shows initially correct, the B. B. K. A. however, despite much criticism, and many shows, had still not got it close to correct. As usual the attitude was one of arrogance – as a group of "Gentlemen" freely giving our time and effort, we cannot possibly be wrong.

Abbott was probably aware of what was about to happen when he made the initial attack in October 1882, and realised that this was to be one of his last chances to express his real feelings. However, whether he, or perhaps the judges at the recent exhibition, knew that Cowan would now withdraw from showing competitively, is unclear. The apparent acknowledgement of criticism is completely out of character.

The November issue of the *B. B. J.* contained the news that from the January 1883 number, Peel would be the new owner and editor. In a somewhat confused editorial, Abbott gives the history of his magazine and it's turbulent association with the B. B. K. A., concluding by stating that he now intended writing a book, which sadly never materialised. The remainder of this edition was full of anti establishment material.

The report of the October committee meeting contained information on four points raised by Rev. Burkitt re shows and the regulating thereof, all of which were effectively 'kicked into the long grass', which he appeared to accept. Abbott was not so conciliatory, pointing out that elections for the Committee were about to take place and voters should question the candidates on the four points. He went on to say that the Committee were fast becoming irresponsible and adding "bee-keeping has become too large a subject in this country to be longer a matter of child's play". He alleges that when he was a member of the

Committee he was asked to go through the show regulations very thoroughly, and make suggestions. All his suggestions were ignored. He believed judges should defend their decisions when criticised, but the terms of competitions have often little attention or consideration by judges when making awards, resulting in some of the errors that he had witnessed. He published many letters critical of judging.

The December 1882 edition marked the end of Abbott's ownership and editorial control of the *B. B. J.* He had eventually bowed to the pressure from Peel and Cowan, and with no possibility of either Cowan or the B. B. K. A. having enough money to tempt Abbott, the magazine transferred to Peel, £2,000 passing the other way. Abbott's relationship with many of the Association Committee members had been difficult after the very early years, and he had become increasingly more frustrated and disillusioned with the relative inactivity, lack of achievement in it's main aims, and, as he saw it, incorrect decisions. Although they shared the same aim – to bring modern beekeeping to cottagers, there was now a fundamental difference between Abbott and Cowan. Abbott was of a strong commercial mind and believed that the B. B. K. A. should be directly involved with working class bee-keepers, whilst Cowan regarded the Association as a philanthropic organisation, and often used that as justification for the Committee acting in an arrogant manner. The level of settlement reflects the differing standpoints; Abbott sold a nine-year old magazine with limited circulation for an amount well above its market value, but had lost an outlet for his views and free advertising for his business.

Abbott had written a letter to the B. B. K. A. that was considered at the November Committee meeting and reported in the December number of the *B. B. J.* As was usual practice, the content of the letter was not made public, but Cowan wrote a response to Abbott, pointing out that he had voluntarily resigned his position on the Committee, thus relinquishing the opportunity to exert his influence on matters. He continued pointing out that the Committee meet the members twice a year and their representatives at four quarterly meetings, and therefore respect and represent the wishes of the members. The incongruity of these remarks must have been recognised by many at the time, and whilst Abbott had agreed not to publish his letter, he seized the opportunity to respond, listing the six points probably contained in the original letter: -

1. He points out that the Committee continue to be lax in the rules for showing – equipment that is never sold for the price given at the show – allowing the same equipment to win year after year.
2. The Committee refuse to lay down the means of spacing frames or their distance apart. The Abbott frame spacing idea is the best, but one

member of the committee, said he would cut the ends off if that became the standard. Abbott's argument is that it helps the cottager.

3. Having pointed out errors in Modern Bee-keeping they were ignored by the Committee. Basically they cannot admit to being wrong.

4. The erroneous judging is a clear indication that judges need to be directed – only improvements to be judged – guidelines to be drawn up. This is the responsibility of the B. B. K. A.

5. At meetings where the members are present the committee normally outnumber them.

6. Committee chosen from those that subscribe 20/- per annum – not likely to produce the best possible representation of the members.

He finished by saying the remarks were in the interest of beekeeping and realised that he ran some risk, acknowledging that he was aware that this would ostracise him.

Following a testimonial fund for him, launched in January, at the quarterly meeting in April 1883, Abbott was presented with a black marble dining room clock, Cowan was not present due to the "serious illness" of Mrs Cowan.

Charles Nash Abbott was 53 when he sold the *B. B. J.* His business was employing twenty five men, had five saw benches, one planer, five foundation machines; he was importing jars from Germany and sections from America. But he had been diagnosed with diabetes and three years later he formally transferred his business to two of his sons. He died in 1894 three weeks after the B. B. K. A. made him an Honorary Member. This was a very belated recognition that any position and prestige they had would not have been possible without his initial work.

The change of style of the *B. B. J.* under Peels editorship was immediately evident. It was to be primarily the Associations magazine, reporting what it had done, was doing or going to do, recruiting, seeking donations, cajoling, rebuking criticism and, elevating the senior Committee members. But much of Peels input had more the appearance of one of his sermons than an editorial. A new regular feature was to be *Beekeepers at Home*. There would be no surprise to discover the first was T. W. Cowan. It was clear that the subjects provided the basis for the article, if not the entirety, themselves. In places it is confused, and as would be expected only includes the information Cowan wishes to be known, and often then only vaguely. For example "…embraced the profession of an engineer but circumstances made it unnecessary to follow that profession". This short biography gives no real information or insight of the individual, but has been used by many, as the basis for numerous articles.

At last the B. B. K. A. began to address the issue of how best to induce the cottager to move to a more humane, modern way of keeping bees, the very purpose for it's existence. It was suggested that *Modern Bee-keeping* be expanded to include a chapter on skeps because there was little chance that on a wage of 10/- to 15/- per week, the cottager, for whom the publication was intended, would or could pay 10/- to £1 for a hive. Papers on the best way ahead were read by Peel and Burkitt, both agreeing that the cottager needed to be educated on how to keep bees in the skep properly first and then move them onto the moveable frame hive. Cowan contributed to the papers saying that he had been trying for 10 years (since 1871), to get cottagers interested in bar frame hives. It was necessary for the Association to "get hold" of a few intelligent cottagers and get them converted to inculcate the rest, because the only way for the advancement by "these people" is by example. Whilst his language would be considered intemperate and demeaning today, it was common in the nineteenth century, but the difference in tone between Peel and Cowan surprisingly indicates that Peel had a better understanding of the plight of the working class. Cowan was not best pleased with the idea of expanding *Modern Bee-keeping* to include more on the skep, and when the motion was put he strongly opposed it, counter proposing that *Modern Beekeeping* be sold in six seperate parts, the skep forming a short chapter. It did not happen but a penny leaflet appeared on skeps in August 1883 - author Cowan.

The B. B. K. A. continued to grow, and after ten years as a monthly the *B. B. J.* became a fortnightly launched with an article by Cowan on queen rearing, and another by Blow on the Anglo-Cyprian hive, probably requested by Peel. As Abbott found, this provides good copy and often much follow-up discussion, the authors responding to criticism, all helping with the additional column inches now to be filled. Cowan was a regular contributor of both articles and comments on other submitted items. He covered the complete spectrum of bee-keeping, often dismissive of other peoples contributions. His contribution to the Blow hive paper, was typical – it is not new, and has problems.

The B. B. K. A. show at Knightsbridge did not draw the public. Cowan was not named as a judge, instruction giver, or prize-winner, giving rise to speculation that he was away at the time. He exhibited his extractors, honey ripener, a solar wax extractor and a device for securing the queen when driving without looking. It would have been very educational if the latter exhibit had been accompanied by an explanation. They were not entered in competition. Had the criticism had an impact? The Knightsbridge event and the show at Bridgwater, held in conjunction with the Bath and West show, at which Cowan was a judge, were both dismal failures, suffering large financial losses, lack of interest from the general public, and would lead to a review of future show policy. The losses of £150 were covered by President (£50 at zero interest) and Committee (£10 per member at 5% interest). Peel appealed for members to increase their subscriptions. The York show at which Cowan also judged faired better.

Having written the leaflet on skeps, Cowan launched himself into a new campaign. He claimed that much of American honey being sold in London was adulterated, and he wanted samples to be tested and prosecutions brought. He did not make it clear how he was able to level such an accusation, but the nature of the statement indicates that he had no means of proving the claim. Cowan was very anti anything non-British; a common position for those of his class at the time and considered very patriotic. This was the start of a campaign to which Cowan would devote much time and energy.

There is evidence that Cowan had been to Germany when in engineering, and visited Paris with two committee members for the 1878 exhibition. In1883 he spent ten days in Switzerland, nearly a week of which was spent at the Congress of Swiss Bee-keepers in Zurich. He reported his experience at the congress extensively in the *B. B. J.* indicating that this was a much more comprehensive effort than the B. B. K. A. had ever considered organising, encompassing genuine scientific research both in presentations and demonstrations, in addition to equipment and honey. As would be expected he is not very complimentary on many things; German hives opened at the back and whilst this enabled them to be stacked, inspection was difficult: French hives were like those of Britain but with frames 14 ½" high and 11 1/8" wide – wrong shape and too large for our country. He did not think much of the extractors – too heavy, frames held in with ropes, very expensive, discarded in Britain as poor:

    The Swiss extract at the end of the season and do not
    extract combs with brood in so they do not require such
    precision as demanded in UK.

Never likely to make a decent watch then!

He considered his extractors to be superior and believed that the Swiss solar wax extractor would not suit in Britain, and that his as exhibited at the recent London show was better. In Swiss exhibitions of honey, more than half the exhibitors got prizes but of small value. He thinks that the B. B. K. A. system of few, high value prizes to be better. He believed it was generally thought that the Germans were ahead of Britain in scientific beekeeping:

    ...so it was therefore, with great astonishment, that I heard
    them express a doubt as to the cappings of brood consisting
    of wax and pollen, a fact admitted by us. M. Blatt has since

the meeting been making experiments and admits that he finds
in my guide book to be correct, and cannot understand how
it is they should not have discovered it before, but had to
learn it from the English.

It is true that the *Guide Book*, when talking about pollen, contains the statement "A small portion is used….for mixing with wax for the queen cells and capping brood cells". Hence the statement that the Germans had to learn it from the English could be correct, but it would have lost it's impact if he had accurately added "…who learnt it from the…..".

When not trying to score points his report on the tour in Switzerland is educational, comparing climate, bee forage bee management techniques etc.

In 1883 Cowan was omni present at all B. B. K. A. meetings until mid July then failing to make an appearance until 19th December. During 1884, when a greater emphasis was being put on Committee attendance, he did not attend at all except for the November meeting. Mid 1884 it was written "Mr. Cowan, though now residing some distance from England". His contributions to papers read at conversaziones were in the form of letters to the *Journal,* some time after the event, only one gave an address – Zurich, October 1884. Careful reading of his account "Amongst the Swiss Beekeepers" and his contribution to a conversazione much later, lead me to conclude that he was in Switzerland for the later part of 1883 and most of 1884 because of his wife's poor health. He would have only needed to return to Britain to deal with his forty-acre farm, and was rarely in evidence at any B. B. K. A. function or meeting. Switzerland was to become a favourite destination for the Cowans in future years; the Shelly family owned property there.

As was now commonplace following a major London show, there is criticism of the judging. One such letter was a very personal attack by Abbott on Cheshire, one of the judges, and was without doubt a continuation of their, now, antagonism to each other; Cheshire was very unlikely to make a show award to Abbott. Neither Peel nor Cowan had seen the letter prior to it's publication, an indication of how both were isolated from the positions they held in the beekeeping world. Peel stated that he would not have published it if he had seen it, because of the inaccuracies it contained, not to mention the criticism of himself and the B. B. K. A. therein. However, he accepted that the B. B. K. A. had interfered and asked the committee of the exhibition to reconsider the awards, which they did. Cowan clarified the error, the Association did not appoint judges at the show in question, but simply provided three names as requested. Semantics as is often the case in such disputes: it is the usual attitude, and no longer surprising.

Peel was now using the editorial of the *B. B. J.* as the means of a crusade for the B. B. K. A. Nothing is omitted and on one occasion he berates committee members who do not attend meetings, but Cowan is excused because he was "compelled to reside abroad" due to the illness of Mrs. Cowan. It is quite an extraordinary situation, an Association that does little other than hold a Committee meeting once a month, tolerating the Chairman of that Committee being absent for most of the meetings, especially as he was the very individual who was the first to lecture on the need for members of the committee to be in a position to attend regularly.

The Association had been under pressure to become involved in helping beekeepers dispose of their honey, after all, the Council members were the individuals claiming responsibility for increasing the honey now being harvested in Britain. For the B. B. K. A. this was never a reasonable proposition because the Association was always struggling due to lack of money. To serve the perceived need the British Honey Company was formed in November 1884, the directors mostly coming from the prominent Association members, Peel being the driving force. At the initial meeting to consider the proposal, only two voted against. Cowan who had previously stated that he was opposed to the Association getting involved with trade, abstained, rather than vote according to his beliefs and offend Burdett-Coutts and Peel. It was referred to a sub committee.

The British Honey Co. was eventually launched seeking applicants for £20,000, in £1 shares, and was to operate from premises in Columbia Market provided initially at no cost by Mr. Burdett-Coutts. Columbia Market was set up by the Baroness in 1868, in the East End of London as a Market Co-operative Society, to provide good food for the poor, a typical philanthropic gesture, but a highly unsuitable area for a honey company. Peel was Chairman and Cowan a director. An offer from a private financier, thought to be a brother of a London honey dealer, to take 5,000 shares on condition that he could take all loss or gain in the first two years was declined, and eventually only just over 6,000 shares were taken by about 400 investors. Cowan held 200. As might have been predicted, judging by the track record of those leading the venture, it was not well thought out, operated from a poor location, modus operandi severely flawed and was under funded.

Immediately, local dealers advertised similar facilities, often offering higher prices, and avoiding the beekeeper having to, initially, send samples for assessment to London prior to the whole consignment being accepted. The Company tried to expand, selling honey packaging materials and launching a new honey beverage – "Mella", made to Cowan's award winning recipe. At its second A. G. M. it was obvious that the Company was not viable, and it was voluntarily wound up "by reason of its liabilities". A few of the share holders bought the equipment and started the British and Irish Honey and Wax Company, capitalised at £2,000, but that fared no better. Cowan lost considerable money in these investments.

Writing much later, Cowan blamed the failure of the British Honey Company on insufficient throughput of honey, the collapse of the price for honey following two good years, and inadequate capital. On another occasion he said the failure was because beekeepers sold their best honey locally, only sending the poorer material to the Honey Company, the business was "set up with large capital and managed by directors of the highest standing" - of course! Whatever excuse one chooses to believe it was an ill conceived and poorly executed project that demonstrated the lack of business ability of those involved - the same individuals who had an indifferent record with their activities at the B. B. K. A.

Rev. H. R. Peel died on 4th June, 1885. He had attended a meeting of the British Honey Co. on 28th May, but had been suffering from "gout in the head and eyes", and on the previous Sunday he had preached on the uncertainty of life. He shot himself in his study on the morning of the 4th. The inquest was held on site, in the afternoon and his funeral was held the next day. Godfrey attended to represent the B. B. K. A., Baroness Burdett-Coutts and Rev. Bartrum sent wreaths.

The next two issues of the *Journal* contained little of consequence and no real editorial. However, in the edition of August 1st there was a notice that Cowan had accepted the offer to become Hon. editor, we are not informed of the background to this. In the next issue Cowan wrote on his becoming editor that the *B. B. J.* was for:

    Free exchange of opinion on all matters connected with
    bee-keeping will be encouraged; but no personalities or
    anything likely to injure anyone, will be allowed.

A similar expression had been used by both the previous editors, but the second part had never been enforced, and the future would be no different especially when the editor was personally involved. In 1886 it was to be issued as a weekly.

Cowan's attendance at Association meetings etc. for 1885 was sporadic, with no distinct pattern. He seems to appear when it is important or beneficial to him. Again he spent much of the year based in Lausanne, but did considerable travelling. It is known that he stayed at Ballaigues in the Jura Mountains and from there he travelled by train, then carriage, over the Simplon Pass ("one of our favourites") to Ornavasso to visit Dr Bianchetti an Italian beekeeper. Eduard Bertrand, a Swiss bee-keeper is known to have travelled with the Cowans, and they all went to the apicultural conference and exhibition in Milan.

On the 6th January 1886 Cowan became Proprietor and Editor of the *B. B. J.* but unlike the transfer Abbott to Peel, details of the change of ownership were not made public, but I doubt if Cowan paid anything like the amount that Peel paid Abbott, or if he parted with any money at all. It was not going to affect the publication greatly, but the sermons would be shorter and subscribers would be spared the Peel family lectures on free trade.

Cowan's attendance at B. B. K. A. meetings in 1886 continued in the same manner as in the previous year. He was present at the A. G. M. and a committee meeting in February, both close to the death of his father on the 1st of February. He had probably travelled from Lausanne for his father and attended the meetings at the same visit. In early Spring he was back in Europe, spending "some weeks" on the Italian lakes, a few days at Pallanza, then to Ornavasso again. During his visit to Dr. Bianchetti on this occasion he was able to study the transition (worker to drone) cells. The change was seldom made in one row normally taking four to six rows. He took impressions direct from combs showing variations, but as usual does not say how this was achieved. He proved that bees can build combs with angles less than 100 degrees. "We found Langstroth's diagram not so very far wrong after all", something he had previously doubted in one of his conversazione contributions. As with most of his experimental work there is no detail on exactly how – just that he did it and this was the answer. They arrived back in Lausanne via the Simplon pass on the 8th June.

He did not reappear in England until 7th July, and between 12th and 16th July he attended the R.A.S. show at Norwich, where he was guide to the visiting Royal party. It was rare for him to visit any Regional shows. He read a paper and was Judge at the London show starting 31st July, where yet again, there were many complaints about the judging.

Later in the year there was a Foreign and Colonial Exhibition at South Kensington to which Canadian bee-keepers brought a large exhibit. A 'conversational meeting' with Canadian beekeepers was arranged for the 20th October, but this was brought forward two weeks because Cowan was not available on that date – back in Switzerland. Many ladies were present at lunch, but not Mrs Cowan who had not travelled from Switzerland. Cowan presided at the meeting, and in his opening address he declared that under the present Queen "Britain's colonial possessions had grown to gigantic power and strength" which was greeted with loud cheers, presumably the Canadians not joining in. He then gave some information on the B. B. K. A. declaring that it had 10,000 members and a staff of experts and assistants of about 50! A. D. Jones responded by inviting any of those present to come to Canada, and followed by describing the Heddon hive. Cowan was not impressed, but he was unusually diplomatic - would not suit in Britain because our climate is more moist. Other contributors to the discussion were also critical of much Jones claimed for America and Canada, but taking their lead from the Chairman, were also diplomatic. The correspondence that followed in the *Journal* was more direct, and persuaded Cowan in an editorial to put the record straight, as he saw it, of course.

He corrected mistakes from correspondents as to attendance, and what they said – or intended to say. He credits the Americans with some inventions but not the hives that the Heddon was developed from, – that was the Stewerton then the Carr-Stewerton. He criticises the inaccuracy of some reports in American magazines:

> The Americans do not, as a rule know much of what is done out of their own country, or we should not have such statements made by an (American) editor as those alluded to last week.

This remark was in respect of Hoge claiming that he supplied American honey to the Queen's table; a matter that was a big irritation to Cowan, because he is convinced that all American honey is adulterated. He goes on to express his surprise that during the meeting on the 20th, at which he was not present, no one responded to Jones when he asked if anyone had tried brood combs spaced less than 1 ½" apart. Cowan had written in the *Journal* many times, that he was against the use of distancing devices, either attached or part of the frame, because it did not permit the frames to be placed at variable distances, particularly at 1 ¼" early in the season to restrict drone breeding and swarming - followers still not paying attention, then!

Cowan sometimes used a few words of Latin as a heading or when in dispute, presumably to demonstrate his classical education. But much of it does not suffer translation or are simple well-used sayings. In the early part of 1887 when he was still away on the Continent he was challenged over *Mel sapit omnia* which he had used several times. His response, when it did eventually come, was waffle.

In the same year, Samuel Simmins yet again launched a new business, this time it was queen rearing. He wrote in the *B. B. J.* copiously on the characteristics of the various races, whilst the same issues carried full-page advertisements for the queens. The declared policy of the magazine did not allow this, but there was no interference, leading to the conclusion that Cowan had a financial interest in the latest Simmins venture. The belief is reinforced when note is taken of the tone of Cowan's report of the demise of the company a short time later – very acid, very close to accusing Simmins of pocketing the invested capital.

The Cowans were back in Britain April 1887, long enough for Thomas to attend a couple of Association meetings and issue the eighth edition of the *Guide Book* before they departed for America, arriving New York 18th July, when they found it too hot for their liking. The main purpose of the trip was to return the visit of the Canadian beekeepers to Britain the previous year, but he had arranged to visit all the bee men he considered important in America en route to and from Canada. It was a very long tour on which he transported his binocular microscope demonstrating it to all he met. He took one of A. J. Cook's classes at Michigan State Agriculture College, and declared "Cook a charming and sterling man" who was suitably impressed with the apparatus. Cowan took great delight in reporting that it was the most powerful microscope ever seen in America, much admired whenever he showed it. The highlight of the visit was to the Toronto Exhibition where Cowan was presented with an address and an engraved gold topped walking stick that was given to the B. B. K. A. at a later date.

T. W. Cowan with his microscope in America

In his response to the presentation Cowan after the usual pleasantries, and his opinion of the exhibits on show, responded:

> The B. B. K. A. is a philanthropic society whose desire was to get as many cottagers and others in moderate circumstances into the business as possible – they wanted everybody to keep bees – the society was formed as a means of bettering the condition of the working classes. The Association distributed thousands of circulars explaining the uses of honey as food. ….. The Canadian system of selling 'honey on a stick' did not commend itself to him. England would not copy in that respect. The B. B. K. A. endeavoured to get as much wealth and as many great people in their ranks as possible. The President of the Association was the Baroness Burdett Coutts, a lady far famed for her philanthropy and liberality. The Vice-Presidents were members of the Royal family and principle nobility in the land. ……The Lord Lieutenants of the counties were generally chosen as presidents of the county associations.

He continued with detailed information on county structure etc. The quote illustrates how Cowan saw the Association and it's mission. It is also interesting to note that, not being a member of the Royal family, he clearly saw himself as one of the principal nobility of the land and considered Burdett Coutts to be liberal, which is probably a good gauge of his political beliefs. He disliked the practice at the Conference of cutting a section into four quarters and selling it for 5 cents, and said so, as can be seen in the quotation above – "very messy and un-gentlemanly", but he was informed that it was a good way of shifting a lot of honey. Eating whilst walking in public not part of a Victorian gentleman's behaviour.

Top of walking stick presented to Cowan

Responding, Canadian Allen Pringle said that in bee-keeping he believed that Canada was equal to Britain, but it was more a business there than a hobby as it was in Britain. He did accept however, that they could learn a couple of things from British bee-keeping organisation. Not exactly what Cowan was there to hear.

Young, from Norway was also a guest but not treated with the same respect. Later Cowan pointed out that the Norwegian government paid for Young to attend the Conference, and it might be a good thing for the British government to consider doing the same.

Cowan was obviously impressed by the scale of beekeeping in America and Canada, the size of the equipment suppliers, the forage available for bees, and the extent of bee breeding. He was, however, shocked to discover that American ladies did their own housework, and was totally floored to also find that husbands sometimes helped, and that it was very difficult to hire servants. He found the accommodation there poorer than in Britain, and in some of the out of the way places they slept on the floor because of "unwelcome bed fellows". Yet another example of the Cowan's privileged life-style, making it unlikely that he could lead the Association on it's declared mission. He had no real experience of those outside of his class, other than his servants.

At a Conference of American beekeepers after the Cowans had left for Britain, one of the proposals was for an Association of all American and Canadian states. Cowan wrote that the greater part of the proposals are the same as those of the B. B. K. A. which he explained to Newman when he was in Chicago, and wrote down for him, except that they have included a Honey Company. It was not passed, but put into the hands of a committee, Cowan expressing the hope that it:

> ...may be as successful and as useful as ours has been ....
> it will of course take some time before it can equal ours
> in efficiency.

The committee put back the proposal by a year before the Conference ended. Cook addressed the conference praising Cowan for his recent visit, but it was very vague, and then made him an honorary member of the North American Bee-keepers Society.

The central association's A. G. M. in February 1888 proposed that there should be one vote per member, rather than the present system, where the number of votes an individual has is linked to the level of his subscription. Something similar had been put forward the previous year. Cowan was not present but wrote:

.... I am quite sure this will be a wrong step to take, and
would make it infinitely more difficult to get subscriptions
than now.   As long as I have anything to do with this
association, it must retain its philanthropic character.
As soon as it becomes something different I shall be obliged
to leave it; it must be a society for promoting bee-keeping
and nothing else.

This quite extraordinary out-burst, appears to have gone unchallenged. The letter implies that members only subscribed more than the minimum in order to have a greater input into the election of the committee.   How would making it a democratic association affect its mission? It might, of course, result in him not being elected Chairman. It was defeated by those who would loose control if they voted otherwise.   There was also the not unreasonable proposition, that anyone who had sold or assigned the right of manufacture could not be appointed to judge at shows. Discussion was in favour until Raynor pointed out that this would mean that Cowan and himself would be ruled out - discussion ends, motion defeated. The question has to be – why is he so concerned? He appears, in this year, to have only attended and chaired one committee meeting and judged at one show, so the Association does not occupy a great amount of his time, and what little it was achieving was without him. It could be argued that he was treating it with contempt.

In an attempt to discredit imported honey to benefit, as he saw it, British honey, Cowan continuously claimed that the American product was adulterated with glucose and latched onto any report that supported his opinion. First a vicar reported that he had witnessed it happening. Published in *B. B. J.* with glee. Newman, described by Cowan previously as a friend, responded saying that it was a lie, and no way should it have been published. Unfortunately for Cowan and the loyal *B. B. J.* followers, when challenged, the vicar could not remember where or when he had seen it, and the originator of the story, says that it was a joke, definitely not appreciated by the target. None of this encouraged Cowan to back away, and he published an editorial, stating that Thurber was the first to import the adulterated product into Britain, through Hoge, his agent, who then started out on his own supplying it and claiming the Queen used it. His Company was now in liquidation; someone else Cowan destroyed in pursuing his view of bee-keeping and his country. As always he provides no evidence to support his claim. He had been made to look stupid, so was not likely to back away, and would raise it at every possible opportunity.

In the same year a similar attitude was exhibited towards Otto Hehner, a chemist and ally in the adulteration controversy, who had stated in a conversazione paper, that wax could be pure white, contrary to Cowan's belief. In a very long response that gave much historical detail but no real justification of his position against that of Hehner, he concludes that he is correct, and has it on a very high and well known authority; so well known that he does not name or quote him.

Mid 1887 the Irish Beekeepers Association had decided upon a standard hive. The frame was the same as the British standard except that the top bar was 16" long and to be spaced using screw eyes, a method that allowed the distance between frames to be varied. The hive was 22" x 14 ½" x 9 1/8" deep; all inside measurements. Timber to be ¾" thick and there was a strip to cover the ends of the frames.

Irish B K A hive 1887

The criterion was that it had to be capable of being made cheaply. Choosing the same basic frame dimensions meant that the same size foundation as the British could be used, but by lengthening the top bar the hive was cheaper to make – no rebate. Cowan was not pleased and stated so in no uncertain terms, he did not approve of the longer top bar; there being a means of spacing the frames; it being a single walled construction; and the brood box being 9 1/8" deep. The British hive was fixed at 8 7/8" deep so that it could be made from a 9" board, planed. Henry Chenevix, Chairman of the I. B. K. A., responded that it was impossible to plane a 9" board to 8 7/8" as Cowan contends especially if shrinkage is to be accounted for, and as to the frame he points out how the 16" top bar was better given the criterion of cheap hive construction. He also objected to a paragraph in which it was stated "….did not adopt the same size frame as the British Standard….". Cowan responded to the latter by saying that it should have read "…..did not adopt the same size as the British Standard.."

and claiming that it was a printers error - so the printers are now adding words! He went on to claim that British manufacturers had no problem with the 9" board because they use good materials. He finished by saying that the thing that we really objected to in the Irish hive was the fixed floorboard - but of course 'we' have only just thought of that. There was no response from Chenevix or none that was published in the *B. B. J.*

I include details of this very petty row because again it provides an insight into the character of my subject. He had decided on the best size for the standard frame, that it should not have spacing devices, and that there was little point in having a standard hive, single wall hives were not suitable for this country, and he had pointed out that a fixed floor was a definite mistake. Furthermore, the B. B. K. A. had, at a considerable financial loss, sent two experts to Ireland to teach the Irish beekeeper modern techniques of the craft, and now they appear to be thinking for themselves! In reality the hive was a very good option for the cottager moving to frame hives, the sole purpose of the design, and not unlike a hive offered by Abbott Brothers. Also, because the legs were an integral part of the hive, there would be no problem making the front and sides from more than one piece of timber; this probably reduced the price further. It would not be the last in Cowan's antagonism towards the Irish, but typical for upper class Englishmen at this time.

At the first quarterly conversazione of 1889, Cowan delivered a talk on *The Choice of a Hive* during which he said, "Although I prefer a loose floor-board, there are others who like it fixed to the hive, but this is not an important point." It seemed to have been an important point a mere six months earlier. He stressed again that he did not agree with fixed distancing methods, and when judging hives he ignored them, something that continually brought him into conflict with Abbott. He states that whilst he was alone for many years in advocating varying the frame to frame distance, depending upon time of year and conditions, since the method has been introduced in connection with the Heddon hive system shown at the London conference by the Canadian bee-keepers, "a good many have followed my example". At the conference he criticised the Heddon hive. In this paper he also claimed to be the first to use perforated excluder zinc on a large scale, (1874), and the first to introduce zinc with 5/24" perforations, but he now does not use any excluder.

In his paper on hives, there are disclosures that give an insight into his bee-keeping. At one time he had forty hives affected with foul brood and had difficulty in eradicating the disease, only getting rid of it when he took the hives to pieces and boiled every part separately. This was with his initial double walled hive, but after this experience he redesigned it so that it was easily dismantled. The only time that he had forty hives was at Comptons Lea. He also claimed to have experimented with artificial heat for hives, and in a direct parallel with Root in America he had tried hives in a greenhouse but found that heated bricks and hot water bottles placed over the supers to be very good adding that they should not be applied before the beginning of March. This is not the best position to place any heating medium, and would have been very labour intensive. He found putting manure around the hives advanced them by a month; stealing an idea from the garden. As usual there is no great amount of detail, unlike Root, who shared every trial and tribulation with his readers.

Again Cowan's Association attendance for 1888 was virtually non-existent but he attempted to maintain his influence through his magazine. It improved marginally the next year, which witnessed a major change in the structure of the B. B. K. A. Numbers were declining, affiliated bodies complaining that they had no representation at the central body. The basic subscription was at last reduced to 5/- and each affiliated body would have a seat and voting rights on the Committee. Cowan had voiced his dissent in the *B. B. J.* prior to the A. G. M. but was ignored, a new experience for him, and an indication that his continuing absence was resulting in a loss of total control. Two *B. B. J.* issues later his editorial carried his displeasure, and gave his version of how it would work, followed by many dictates - decidedly unhappy. However none of this would solve the Associations financial woes. By the end of 1889 it was again running out of money, but it is difficult to see why. Of the reasons given, only show losses are credible, unless large expenses are involved.

Cowan returned from Lausanne to be at the R. A. S. Jubilee show at Windsor, held during the last week in June, 1889, and was present all week. The Queen made a cursory visit to the bee department where she was met by Burdett-Coutts and Cowan who in his report was clearly delighted with being presented to the Queen, but unable to conceal his disappointment that her visit was nothing more than courtesy, staying the minimum time. He had taken time off from his holiday, travelled from Switzerland, remained all week, and the thank you for the gift given to the Queen, came from one Henry Ponsonby, the Queen's private secretary, and very curt at that. Again, my subject not pleased, but you would not know it when reading his effusive report on this event in his Jubilee history.

His attendance at B. B. K. A. was marginally improved in 1889 on the previous year, but in most respects this was an uneventful year for Cowan, the impression gained is that of someone beginning to loose interest.

Langstroth passed the rights of his book to Charles Dadant and Son, who revised it and issued their first edition in 1889. A feature was the full plate portraits of well-known beekeepers. Cowan supplied his as shown; he was in his late forties. The caption simply read "J. COWAN. Editor of the *British Bee Journal*". He was not amused and wrote to the Dadants insisting that the caption be corrected in accordance with the information that he had supplied, and an apology be inserted in the next edition of the book and immediately in their magazine. Cowan had cultivated the image of an internationally well-known and respected authority on bees, especially on the other side of the Atlantic. His critics were quick to point out that he was clearly not that well known. With the revision of the book came a revised caption: THOMAS W. COWAN, F. G. S., F. R. M. S. Editor of the *British Bee-keepers Guide* and publisher of the *British Bee Journal*. Cowan was so incensed that when next in America he made a considerable detour to visit Dadant, but there was no published apology.

Cowan portrait from *Langstroth on the Hive and the Honey Bee*, Dadaant edition 1890.

1890 was of the same pattern, appearances early and late in the year, away in Switzerland for Spring and Summer, where he once again toured beekeepers, this time in France, reporting back to the Association faithful late in the year. However, the year was not without controversy. In the very first issue of his magazine for this year he indulged in a little self-praise, detailing how well the Association and the *B. B. J.* were doing under his leadership, reiterating his mission to exclude all 'personalities' from his magazine. He followed with a letter from an un-named correspondent congratulating him on the improvement in the *B. B. J.* since he has had it, "…keeping out the personalities correspondents indulged in formally". What magazine Cowan was referring to or his correspondent was reading, bore little relation to the one published. Given that it was a beekeeping magazine for bee-keepers by bee-keepers, complete unanimity was impossible.

As if on cue, the disagreement that appliance manufacturers/dealers had been having with the B. B. K. A. over their ban from being members of the committee, (Cowan and Raynor excluded because they were not manufacturers), took a turn for the worst. In protest at the committee's intransigence in the matter, Abbott Bros., George Neighbour & Sons, and Thomas B. Blow had agreed not to exhibit at the R. A. S. Plymouth show. Needless to say, Cowan was not pleased and said so in an editorial, but put the failure of the show down to many factors, and the manufacturers were absent because they were very busy. Thomas B. Blow responded with what can only be described as a rant, using expressions such as "…tyranny of the Committee of the B. B. K. A." and then having said that he did not wish to go into details of why the manufacturers felt aggrieved, he does just that, pointing out that they had not been too busy in previous years to exhibit. He relates the instance when he should have been elected to the Committee but was thwarted by "…what was at least a bit of sharp practice on the spot", after which they changed the rules to eliminate the possibility of a manufacturer applying for Committee membership, - Cowan and the Committee at their arrogant best. In order to maintain balance there followed letters supporting the Associations' stance, and from dealers distancing themselves from Blows outburst, but complaining none the less, about the *B. B. J.* selling items for beekeeping, and the deposit scheme that it had launched which none of the dealers were prepared to take part in. Cowan expressed surprise that dealers would not participate in this scheme, which in itself is surprising, because it added to their costs whilst questioning their integrity. Not wise for a publisher to upset his advertisers, but they had nowhere else to go. He tried to exit gracefully saying that the deposit scheme was really aimed at the private advertisers, for which it was an excellent scheme.

On bee-keeping matters Cowan stated that he now prefers the English black bee and all others should be left alone. This follows many years of experimentation with other races, and presumably means that he no longer holds the Italian bee in such high esteem as he did previously, as he stated in the discussion on Raynor's paper "The Ligurian Queen Bee", in 1879. Whilst he had written much on the different races of bees, there is no indication as to the reason for his change of mind. He reappeared from his annual sojourn in Switzerland in time for the October quarterly meeting where he was asked to:

> ....kindly oblige them with a recital of his observations and experiences respecting beekeeping gained during a recent visit to some of the south-eastern departments of France.

The Chairman said that he had not prepared a talk but did have his note-book and some photographs with him!

> Thoroughly broken down in health we left England (in May), to try and regain some strength in our favourite country, Switzerland.

Whilst relaxing in Switzerland he had been invited by E. Bertrand to join him and De Layens both of whom were also taking a rest from their labours, on a bee-keeping expedition. It was a very long tour, visiting nineteen apiaries, and examining several hundred hives. Most kept their bees in Layens hives, but the few that had Cowan hives were given special mention, as were those who used the *Guide Book*. The trio were treated like bee-keeping royalty, he was very pleased, and promised more in the *B. B. J.* later. During the discussion he disclosed that he had been to an apiary stocked with Cyprians and Cyprians crossed with blacks and found them perfectly quiet "thereby disturbing English notions concerning hybrids", especially, of course, his own frequently expressed views.

In the magazine of the following year he carried out his promise to give more details of his beekeeping tour in France with de Layens and Bertrand, and it is in this report that he gives the impression that they had a property in Switzerland. I believe he leased or used a Shelly family property, but although the address is known, the Swiss authorities have been unable to provide confirmation.

Recounting the invitation from Bertrand, Cowan wrote:

> ......we had already made a pleasant trip together in 1883 ....we could do this with a change of scene as well as by staying at home – which was always a temptation to be at some sort of work, which was strictly prohibited.

They were there to recover from the 'flu.

During the year there had been a build up of tension between Cowan and John Hewitt, although there was no evidence yet of the escalation. Towards the end of the year, Cowan's fight with the sellers of 'liquid fruit sugar' escalated. D. A. Hume & Co. had sent a sample of their product to the Committee of the

Lancashire and Cheshire B. K. A. (and probably others), seeking their opinion on it for feeding bees. The accompanying letter stated that it was made from beetroot in Germany where it was widely used to feed bees and attached the composition. Some County Associations sent it on to the *B. B. J.* seeking their opinion, rather than ask the central Association. The response from Cowan occupied a complete column, summarised thus: it was originally brought out by Dr. Follenius with name 'fruit sugar', it is introduced under a false name – it is made from beetroot. He then quoted Dr. de Planta, naturally, his opinion supported Cowan's view. Even on the continent where sugar is at least twice the cost in Britain, it did not have universal support among beekeepers and hence Cowan saw no reason for its use. He also pointed out that if, as it was claimed the syrup was made from sugar, it could only be cheaper if it was bulked out by less expensive additions, although he did not name it – glucose. There was not always logic in Cowan's outbursts. He added:

> It is not surprising that the manufacturers of this liquid syrup can find beekeepers to give testimonials. Every quack medicine vendor can get his testimonials.

Cowan's problem with the product goes back to the honey adulteration scandal, and he believes bee-keepers should not handle anything associated with glucose, because the public might believe they would be using it for unlawful purposes. "On all grounds we should condemn the use of this so-called 'liquid fruit sugar'." He ends by casting doubt on the ability of the substance to keep bees healthy or even alive.

One month later the editorial of four pages returns to the subject of 'liquid fruit sugar'. The editor initially defends his right, "our duty" to express the views and publishes a letter from the agents for the material:

> ......in order to state still more forcibly the fatal mistake we consider bee-keepers will make if they give countenance to this so called liquid fruit sugar in any form or shape whatever.

No holding back then. The agent's letter was well written, and points out that Dr. de Planta had retracted the remarks quoted in the *B. B. J.* when he addressed Swiss bee-keepers. They then addressed Cowan's remarks on the name, testimonials and "unlawful use" and concluded by requesting that the letter be published:

> ..... so as to remove any unfavourable impression which may have been formed amongst the readers of your *Journal*, through your imperfect knowledge of, and unwarranted strictures on,

liquid fruit sugar.

There followed further communications, Cowan would never back away, but much is a matter of semantics, Cowan attempting to rename substances to fit his argument. He then answers at great length every point – again. Dr. de Planta was deceived by the name apparently! The originator of the substance had said, "fruit sugar is honey without the aroma". Cowan responded to this with:

> We most emphatically deny this, and claim that the aroma
> is one of the principal constituents of honey, and that no
> compound without aroma can be honey.

It was, of coarse, not really what the agent was saying. I have omitted the details of the chemical analysis that formed part of this argument because it was of its time, and not conclusive. Cowan's difficulty was that he confused feeding bees and honey adulteration. No matter what liquid fruit sugar in reality was, if it sustained the bees, and bee-keepers were not adding it to honey to sell as honey, there could be no problem. The only real solution that would satisfy his argument was to ban glucose from sale in the country. In reality Cowan had responded hastily in his characteristically arrogant manner, to a reasonable request, and thought the matter would end there, illustrating his lack of experience or knowledge of life outside his very restricted circle. What evidence did he have for his stance? He never withdrew from any argument, no matter what the circumstances or facts, often, as he did on this occasion, insulting someone whose expertise he had previously lauded, and often introduced 'red herrings'. He received a further communication in the New Year from the agent but he declined to publish it – "nothing new". Since when had that been a drawback in his magazines?

We can only guess at what happened next, but it is my opinion that the agent for liquid fruit sugar did not let it lie, and legally sought damages. The justification for this view is that by May 1891 he again ran an editorial warning against using beet sugar for feeding bees, stating that only pure cane sugar would do, and because it was difficult to source, (he had only just found out!), he was to sell it at cost, from the offices of the *B. B. J.* So now he further antagonised the dealers who sold cane sugar for profit, whilst providing an income stream for his magazines – at cost did not mean without profit! Twenty percent had been added to 'cost'. Another ill thought out incident.

On the domestic front, 1890 saw a major change for the Cowan family. For most of the year it followed the now established pattern, based at Comptons Lea, but with most of Spring, Summer and Autumn spent in Switzerland. By

October they were back in Britain and residing at 31, Belsize Gardens, London. The Cowans no longer had any need to be in Horsham. Cowan's association with the Michell family business had ceased, he always claimed that he made no money from the forty acres at Comptons Lea, and they had not spent the greater part of the last seven years there. Their children were from twenty one to fourteen years of age at the time of the move and all still living at home. Thomas and Fanny were both now fifty years of age, and London was more convenient for his activities. He did not appear to have kept bees at Belsize Gardens, the few that he now kept were housed at a property he leased in Cornwall.

Belsize Park Gardens was described at the time as for the "upper middle and middle-class wealthy", houses costing between £100 and £130 p. a. to lease.

Present day photo of adjacent house in Belsize Park Gardens

The change in abode appears to have had an adverse effect on Cowan's temperament, becoming even more truculent. Following his destruction of the liquid fruit sugar agents, in an answer to a query about the legality of beekeepers making mead, he responds by making nonsense of the law and says that he will gladly pay any fines levied for such a stupid law. It was obvious from the response that he had no idea of what the law was, and a couple of issues later Otto Hehner (the B. B. K. A.'s analyst), gave the letter of the law. It was not really what Cowan wanted to hear, but he re-iterated that he would pay any fine levied on the original correspondent:

> ......if for no other reason than that it will enable us to inform beekeepers what to expect if they continue to sin in that way.

The promise has been down graded from any beekeeper to the single correspondent. Cowan was a staunch teetotaller so the quote is meant literally, which makes his initial response all the more difficult to understand. Yet again it is a question of instantly responding to a situation, and thinking later. There was no further report on the matter.

He was also incensed by the collapse in a very short time, of the latest business venture in beekeeping by Samuel Simmins. The implication is that he and several of his friends had again lost money in the business. There were debts of nearly £3000, and assets of less than £50, resulting in a first, and probably only, payout of less than ½ p in the pound – "a discrepancy altogether beyond our comprehension".

Next he grappled with foul brood specifically the Canadian treatment of it, and not for the first time. The *B. B. J.* editors had declined to publish a letter from Canada describing the manner of their successful treatment of foul brood, citing that it was not new and they had tried it without success. Directly on cue, "X-Tractor" a barely disguised pseudonym, criticises the Canadian method and praises Cowan's modus operandi for dealing with foul brood. Pringle, who Cowan had befriended on his visit to Canada, responded in very strong terms, leaving him no real option but to publish his letter, heading the article "Foul Brood. The Canadian Treatment: is it new?" Basically the method was – when bees are flying well, in evening, shake them into their own hives and give foundation starters. In the evening of the fourth day remove the comb and give fresh foundation. Put all the diseased combs into a double box hive and seal up for two days; after that open the entrance and when most of the sound brood is hatched, remove those combs and give bees starters of foundation in single box hives. After four days remove the combs and supply bees with new foundation. This works by believing that honey is the mechanism for spreading the disease, and not the queen, wind, or other, as preached by Cowan:

> The fact that our method will cure and invariably cure (so far in our experience), when properly carried out ought to be sufficient for any reasonable man who is not lost in his own conceit.

Not friends now then! The method, attributed to William McEvoy, had been in use for fifteen years (from 1875), and Pringle claims that the reason that it had not been propagated in America and Britain was due to:

```
......prejudice, preconceived opinion, the unwillingness we find
in human nature to accept a new truth or fact, especially
if it comes from an humble or unorthodox source.
```

As would be expected the editorial footnote was long and not complimentary. Della Rossa (should have been Rocca) and Schirach had both tried the starvation method and discarded it over 100 years earlier, followed later by Berlepsch, and Dzierzon. Cowan had tried it fifteen years ago, obviously not taking the word of the bee masters quoted, and it failed. Did not work for him then, - familiar? He naturally objects to the sentences quoted above, although many might believe that they sum up his character well; also the English did not need to be taught beekeeping by the Canadians. He concludes that the foul brood in Canada must be different than that in Europe. Viewing this with twenty-first century knowledge it is easy to pour scorn on both houses, but it was very early in the research that would eventually identify and name the two forms of foul brood in 1907.

In order to turn this Victorian domestic disagreement into a farce it only required John Hewitt to enter, claiming credit for the method proposed by the Canadians. He duly obliged, appearing a couple of issues later doing just that. He had published it in the *Record* in 1887 with one important addition – the disinfection of the old hive. He was sure that honey was a means of infection "no matter what any one may say to the contrary", because of its sticky nature, the spores of the disease adhere to the surface, which is then licked off by the bees, but it is not the <u>only</u> medium for spreading the problem. So his investigation had been scientific. It is the usual Hewitt claim to be the inventor of everything bee keeping, but being denied the credit. Cowan was quite restrained in the foot note – it could not have been foul brood proper (whatever that means), any way, it was published over 100 years ago and in the *B. B. J.* of 1873. I was unable to find the latter reference. I have covered bee disease regulation in detail, elsewhere.

The main show in 1891 for the B. B. K. A. was, as usual by now, part of the R. A. S. show, this year at Doncaster, on 22nd, 23rd, and 24th June. Cowan and Broughton-Carr were in attendance for all three days, our subject as steward, but not judge or exhibitor. The central body also held an 'away day' in July at Wakes Colne rectory, the home of Dr. Bartrum. Despite many mentions in the *B. B. J.* and much organising by Huckle, few attended. The Cowans were both present, fitting in a few days visit at the same time. Given this evidence in conjunction with Cowan's presence at all things beekeeping, one concludes that the residence in Switzerland had been relinquished. Living in London made attendance at all meetings easier, and Mrs. Cowan also attended a conversazione, presumably much recovered in health and able to stay in Britain for most of the year.

The controversy regarding single versus double walled hives occupied many column inches of the *B. B. J.* The County Associations were beginning to behave independently of the central body, particularly by seeking and obtaining funds for beekeeping education from their local County Councils, - money allocated by central government for technical education, which enabled them to appoint their own experts. This was a considerable snub to the Central Association. There is little doubt that the magazines are more interesting when Cowan is 'hands on', but much is confrontational due to his combative character. Working through the issues of 1891, can be compared to watching a volcano, knowing that it will, at some time erupt. There was not long to wait!

Residing in London and Cornwall resulted in Thomas taking a more active part in Association and Journal matters, but 1892 saw him completely consumed by the Hewitt matter and Bee disease legislation, although he did manage to fit in a visit to a bee conference in Switzerland as well as his extended visit to North Africa.

At the beginning of 1893 the *B. B. J.* carried a retrospective, much being devoted to the Punic bee quarrel and included the quite remarkable line "the *B. B. J.* has neither criticised in an unfriendly spirit, nor helped anyone to 'grind their own axe' to the detriment of others". Not applicable to the editors, presumably. However, the tone of the magazine appears to become considerably more conciliatory. Having disagreed with a correspondent about his view on selling honey by volume, when confronted with some basic facts the editors changed their statement claiming an error. Yes, that was conciliatory. Cowan had always contended that bees did not get pollen from heather, but now he adds that they might get pollen, inadvertently when visiting heather. It is the closest he gets to a climb down. Cowan spent two months of the Summer in Switzerland, France and Germany.

Normal service soon returned after Cowan made some less than complimentary remarks about Colonial honey especially that from Australia and New Zealand. During correspondence with an Australian honey producer Cowan claimed that the Australian bee-keeper's eucalyptus honey, of which he had been sent a sample, was not fit for the British table, and the British palette would not take to it, advising that the honey would be best promoted by highlighting it's medicinal properties! Following further unfriendly correspondence Cowan eventually drew a line under it with "we have nothing to qualify or withdraw from what has appeared in these pages with regard to Eucalyptus honey". Mysteriously the honey was confiscated by Customs upon its entry into England. It was Cowan's rather cumbersome attempt to protect the home honey market, and reflected his campaign against American honey claiming that it was adulterated. A year later he repeated the exercise with River Plate honey – "altogether unsuited for the British palate". It is well meaning but completely misguided and arrogant – is he is now the authority on the nation's palate?

Disillusionment with the B. B. K. A. was beginning to be expressed, with several correspondents, yet again, questioning what the Central Association did in return for the contribution from the County Associations. There was, as would be expected, a stout defence, and yet again Cowan stated that the central body was a philanthropic organisation, adding that return for money should not be a consideration. On many occasions he threatened to leave if this was not the position. But the questioning of some of his statements and challenging his views on certain matters began to build. The increasing membership of County Associations brought in individuals who were not prepared to blindly follow the master. A firm of analytical and consulting chemists challenged his dialysis of honey. A bacteriologist suggested that a qualified member of his profession would be required for disease legislation; the B. B. K. A. experts would not suffice. Since being drawn into the Hewitt affair by deceit, the *A. B. J.* had become much less friendly to Cowan, publishing correspondence that previously would not have appeared. It was no longer such plain sailing as it had been.

However, my subject seemed undeterred, re-igniting the Australian and American honey controversies, refusing to publish some correspondence that he did not like, and in typical style, he wrote to Agricultural and Technical Colleges enclosing details of the examination the B. B. K. A. would undertake for applicants of Lecturer in the science of Apiculture now being created. Not that anyone had asked him.

But did he have plans? He did indeed!

Chapter VI

John Hewitt, Dr. Robert Hogg and the
Journal of Horticulture

John (sometimes James) Hewitt, was born into a Yorkshire working class family in 1850. He benefited from the little education that was available to those of his class; became an apprentice in the steel industry; and eventually was employed as a Mark Forger in Sheffield. In his early thirties he married and had one son. In 1874, he set up in business 'on his own account', manufacturing and selling patent glasscutters, which were remarkably similar to those available today.

Each 1/-

Post free, 1/1½.

Each 1/-

Post free, 1/1½.

HEWITT'S PATENT

GLAZIER'S CUTTER,

Invaluable to Gardeners, Amateurs, and Mechanics.

Having developed the usefulness, popularity, and commercial value of the *Wheel* Cutter as a substitute for the diamond in cutting glass, we now introduce the above as THE MOST PERFECT development of this popular tool.

Tempted by our success since 1874, others have tried to imitate our goods, both here and abroad, but as all have (up to the present) failed to make the wheels HARD ENOUGH to *wear*, the disgusted buyers have often unjustly condemned *all* Wheel Cutters.

To prevent disappointment buy only **HEWITT'S**.

Hewitts advertisement for glass cutter

Hewitt had been involved with bees from the age of eighteen, but it was not until 1872 that he had his own. Not content with simply keeping bees, he began to experiment, and as many others were doing at the time, started looking for ways to make money from his hobby.

Although he states later that it was 1883 before he knew of the *B. B. J.*, he contributed to it a year earlier, initially in *Echoes from the Hives* with reports on weather; how bees were performing etc. – just bits and pieces. The same year marked his first major contribution to the magazine.

At a conversazione in April 1881, C. N. Abbott, at the request of Peel, had read a paper to the assembled clique on "The Hive for the Advancing Beekeeper". A large part of this paper, was concerned with methods for spacing the frames, something that had occupied his mind as a manufacturer of hives, and that he was passionate about. It was an ongoing source of disagreement between him and Cowan. Abbott concluded that closed end frames as in the Giotto hive, were the best solution. This hive was first introduced into Britain by Danyell via the columns of *The Field,* and exhibited by Cowan in 1876 at Alexandra Palace, for which he was awarded a prize.

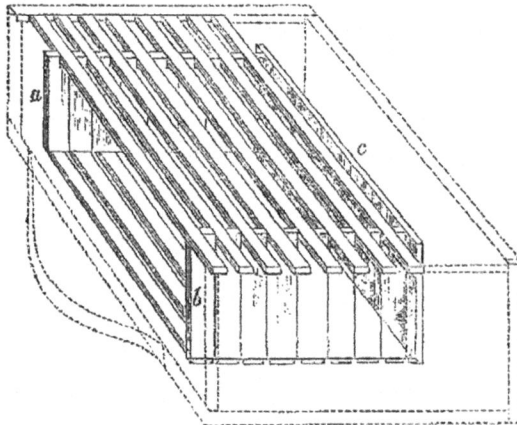

Abbotts version of the Giotto hive

Cowan, who was in the chair, was first with his remarks. He thought propolisation of the end frames would be much more of a problem than with his system – metal runners, Novice's metal frame corners, (which he later abandoned), and no spacers, thus enabling the bee-keeper to vary the frame spacing according to the season. Also closed end frames needed a special extractor. He had tried the Giotto hive, found it troublesome and had given it up; although he had thought highly enough of it at one stage to exhibit it and take the prize awarded. He also said that he preferred to start sections in the body box, moving them to a super for finishing, but did not explain why this was not possible with closed end frames.

Cheshire also took exception to many of Abbott's observations, using the opportunity to continue his feud with Abbott who must have wished that he had stayed at home! The paper gave rise to a substantial amount of correspondence in the *B. B. J.* on the pros and cons of closed end frames, not a surprise to any present day bee-keeper who has ever attended a local Association meeting, where such arguments are rehearsed over and over again. Eventually, up steps Hewitt to have his say. He was, of course, not at the meeting, but joined in with a contribution that did little to add to the discussion or solution, just questioning the ability of an amateur to make the closed end frame; he agreed with Cowan on propolisation. Abbott strongly defended his position bringing a further letter from Hewitt, which was only really a repeat of his initial letter. Obduracy all round!

This little episode provides an insight into the character of Hewitt, a strong, abrasive, self-opinionated individual, not afraid to 'speak his mind' and able to articulate well for someone from his background, if not at the same level of sophistication as the London club members. He saw himself as the natural spokesman for the cottager, but he also saw the possibility of making money from his hobby. However, he could not have believed that he would create such a problem for those at the top of the B. B. K. A. as he eventually did.

Hewitt's next excursion into print indicated where he saw the possibility of commercial gain in the beekeeping world. As with many beekeepers at this time he was experimenting with different races of bees, seeking the Holy Grail of a bee that would produce large amounts of honey, was easy to handle, and not subject to disease. In 1883 he wrote strongly praising the native bee of Syria, claiming that he had been promoting them since 1878 - there is no published evidence of this. He obtained his Syrians from Frank Benton a breeder and exporter of many races of honey-bees. How he achieved this is not clear because it was not until the end of 1884 that Benton wrote that he had discovered that bees could be sent from abroad through the British post simply by registering the package. The water was often not clear when Hewitt was involved.

A Syrian queen from Benton at this time cost 22/-, well beyond cottagers and most other's financial ability, and not, one would have thought, within Hewitts. Initially Hewitt was not involved in trading these queens, but he discovered that some beekeepers were having difficulty in introducing the queens, and he offered to supply a swarm headed by one of Benton's Syrian queens for 42/-, a price he stated would bring him no profit, claiming that, for him this was not a commercial enterprise. At this time typically a swarm headed by a British black queen was 15/-, and with a Ligurian queen, 20/-, so Hewitt's offer was reasonable but not without profit.

It was not long before Hewitt expanded, importing Syrian queen bees from Benton and selling them on, with or without a swarm. He used the well-trodden path of writing long articles in the bee magazines detailing their good characteristics, at the same time placing a small advertisement. However Syrians were generally disliked, even their advocates admitting that they did not like smoke, making them difficult to handle. However it was the first cross with either the Ligurian or English black that was their undoing, the resultant colony being nearly impossible to handle and most beekeepers gave up on them at this stage. C. N. Abbott was one of such beekeepers. He wrote glowingly of them in 1882, saying that he had made £20 in honey sales from a single colony of Syrians in one year. However, two years later he wrote that he had cleared them out of his apiary - "they are the most savage, uncompromising wretches we have ever met with". Hewitt wrote admitting that they were not happy at being smoked, but pointed out that Abbott had changed his mind, and he was probably reporting on hybrids and not the pure breed. It later transpired that the queen Abbott was writing about was probably not a Syrian anyway, although it was given to him as such, having been in the possession of two other beekeepers before he got it. Abbott demanded that Peel, the then owner of the *B. B. J.* publish an apology, believing that Hewitt's remarks were a stain on his character. Peel acceded to the demand, but then refused to publish Hewitt's response. A bizarre event, given that Hewitt was only stating the facts – Abbott *had* changed his mind. It illustrates how the central group would 'close ranks' when under attack.

Hewitt hive showing 'lifts', roof and floor

Hewitt wrote on many beekeeping topics. His articles were directed at cottagers, which was a rarity at the time, even though that was the declared mission of the B. B. K. A. He wrote at great length about his hive, the "Excelsior" giving very detailed instructions on how "any low grade carpenter" could make it. Essentially it was an empty De Kyper brandy box inside a series of lifts very similar to those of the, yet to be designed, WBC hive. R. O. B Manley believed that Hewitt was the first to use these 'lifts'.

Hewitt showed how to modify a standard dessertspoon to give it a lip in order that it could be used for pouring wax to fix foundation to the top bar. He gave details on how to make candy and described its use for feeding in Winter, but was criticised because he was reporting on two hives over one Winter - something that had never bothered others. Much correspondence followed with Hewitt strongly defending his proposition. Comparison with Cowan is inevitable. Peel, probably prompted by Cowan, eventually closed the correspondence and refused to print Hewitt's final letter on the subject; it was the only way he would stop. Hewitt went on to suggest 'improvements' to several aspects of bee management as taught by Cowan, always laying claim to the ownership of the idea - something that he was obsessive about. A specific example being his modification to Cowan's method for queen introduction in the *Guide Book* claiming that:

> ......most beekeepers will marvel at this, but my observations
> and experiments have enabled me to lay down the law, that
> if a hive has no queen, nor means of raising one, they will
> accept a fertile queen at the flight hole.

Later he made a habit of 'laying down the law' often labelling techniques 'Hewitt's law' or 'the Hallamshire law', making Cowan furious. For someone with the personality traits of Hewitt it was inevitable that before long his articles and letters would cease to be solely constructive. Soon he started to directly or indirectly criticise individuals, organisations or both.

Here are some:

- Cowan, for the wrong choice of size and shape for the British Standard frame. He was, of course, not a lone voice in believing that the chosen frame was too small and the wrong shape. Cowan was called upon many times, to defend the shape and size chosen for the B. S. frame, and even though he gave a variety of answers he was never able to quell the opposition. He claimed that he had researched it in Horsham, but much later conceded that it was virtually the frame he used in his own hives. As usual when the decision was made on personal preference rather than logic he reverted to the expression "...all advanced bee-keepers agree with me....". The thinly veiled insult was, of course, never an adequate substitute for a well-argued response

- Cowan, for the wrong size section. Neither the B. B. K. A. nor Cowan had chosen a standard size, but had suggested that only one size would be judged at shows.

- Cowan and others for use of Greek and Latin in some of their articles in the *B. B. J.* etc. Hewitt said that it was not understood by the majority of beekeepers and was just a show of cleverness. Attempting to understand some of the Latin expressions Cowan used also casts doubt on his ability, and on one occasion a brave soul actually questioned something he had used; no result of course. Many were common Latin phrases.

- B. B. K. A. for the badly drawn rules for the Bligh competition. Hewitt was one of many to voice the same complaint but he was the only one to directly accuse Cowan of rigging the result, of which there was substantial evidence, and significantly, drew no response. The solicitor representing the individual who was the obvious winner, but not a member of the inner cabinet of the B. B. K. A. suggested more kindly that the judges had been "unknowingly deceived by plausible misrepresentation".

- Broughton Carr (Cowan's nomination), for his judging at the Doncaster show, at which Cowan appointed himself steward. The accusation was that there was no anonymity. Hewitt claimed that an error in the report by Carr in the *B. B. J.* in respect of a date in the schedule was proof of this. The detail is complicated, but Hewitt was probably correct; it was a perpetual problem at early shows.

- Cowan for his honey ripener – "a toy for those that can afford it, but bees collect in day and ripen at night, so by extracting in the morning removes the need for device." The Cowan teaching on extracting included uncapping first -when the honey would have been ripe.

- Cowan and the B. B. K. A. for seeking legislation to destroy all bees and associated equipment with foul brood. Hewitt claimed it was not necessary and wrote articles describing how to deal with it, but they were refused for publication the reason being that it was the well-known starvation method, which did not work. True, it was only a small modification to that method. I cover more of this aspect in a separate section.

- B. B. K. A. for not giving judges greater guidance at shows. He quotes an example where in a class for observatory hive complete with pure race of bees, first was awarded to a hive with a queen-less colony of hybrids.

- The zig-zag entrances. In itself not a problem, but he included C. N. Abbott with F. Cheshire and others as recommending these devices, which he believed had been responsible for killing many bees. For Abbott at this time, being mentioned with Cheshire in any regard was a major offence because they were now enemies. But Abbott disliked the entrance for the very reason given by Hewitt, and wrote strongly to protest – ignored by Hewitt!

In an article in the *Journal of Horticulture,* April 1885, Hewitt promoted the idea of a gigantic trade union of all British bee-keepers, as a response to the Honey Company being launched by the leaders of the B. B. K. A. and others. It was a reaction to cottagers believing that they would not be treated fairly by the Honey Company and that it would increase foreign competition; exactly the opposite to that intended. In the next *B. B. J.* editorial, Peel responded by delivering a lecture on free enterprise, directly linking it to his support for the Honey Company writing:

> ...... that we British beekeepers have anything to dread from
> foreign competition, we mean as far as the extinction of
> the beekeeping industry is concerned, is, in our opinion as
> remote a contingency as the independence of Ireland.

This was at a time when Prime Minister Gladstone was trying to get a Home Rule for Ireland Bill through Parliament, finally failing in 1886. Ireland was a target for all those at the top of the B. B. K. A. at some time.

However, Hewitt was never easy to deter from a cause and in June 1887 he distributed copies of the prospectus for his Union, which drew the following entry in the *B. B. J.* by Cowan:

> A WORD OF CAUTION. A singular document, purporting to be
> a prospectus of a bee-keepers Union, has been forwarded to
> us. It contains no names either of the promoters or the
> secretary; the name of the printer is also omitted. The
> document is rough in arrangement, slatternly in style,
> ungrammatical in construction. Glowing hopes are dangled
> before the bee-keepers vision; but how they are to be
> realised we are 'lost in conjecture'. Like some cheap
> medicines which profess to be panaceas for all 'the ills
> that fleshis heir to,' the prospectus proposes to accomplish
> everything the bee-keeper can possibly require at little
> or no cost. We note the articles which provide for the
> defraying the expenses of the Council. We advise our
> readers to receive the statements of this prospectus with a
> large 'grain of salt', and to exercise caution before they
> part with their cash.

The prospectus was, of course, not published in any of Cowan's magazines and I have not seen a copy, but the content above begs the question; if the document contained no names, to whom were bee-keepers to send money?

Hewitt also wrote in the *Journal of Horticulture* that the British Honey Company was not the solution to the marketing of cottager's honey, simply stating some very obvious facts, thereby allowing him to crow when it collapsed after a very short time. It was not difficult to predict the problems, but when it was eventually wound-up, it was surprising to discover that it had cost 1/2 to sell 1/- worth of honey, and that with no rent to pay for premises, and no remittance for the directors, allegedly! But as Hewitt was to repeat many times "any lad from a wholesale warehouse could have told them".

Another problem arose when, in 1888, Hewitt took out a patent in Britain for glass sections, (Patent No. 7660, 1888). This outraged many, including T. Bonner-Chambers, Rushton, Nicholson, 'Woodleigh', C. N. Abbott, and of course, Cowan, all of whom laid claim to be the first to invent glass sections, the latter taking the opportunity yet again to express his disgust of patents in beekeeping:

> ....in an article which very probably may be much in request in the future; the more so generally the words of a patent embrace much of the ingenuity of those that have preceded the patent.

It did not seem to concern him when he was running his engineering business.

A letter from Bonner-Chambers at the time of the patent, summed-up the history of the glass section in Britain thus: - he wrote of them 10 years earlier, 1878, and quotes Raitt, (now dead), as having shown some at Dundee in 1878. For presentation a glass back and front was added, Abbott had apparently queried how the bees got in. Raitt claimed Abbott copied the sections of his and a John Stewert of Arbroath shown in 1875. Stewert's were all wood, Raitt's had a glass bottom rail, it was this that Abbott copied. Abbott's were combined into a divisible super. Raitt then moved on to glass on all sides, the top being split to take the foundation; Lee now made these sections. Cowan claimed that his were a year earlier than Stewert, but as is often the case there is no evidence. Every bee-keeper thinks he has invented something!

Prior to Hewitt's patent. the four pieces of glass forming the glass section had been held together by devices at the corners. The patent, that contained 25 pages of drawings, of 75 illustrations, was for a section made up from 4 pieces of glass that were held in position inside a wooden section. The wood was peeled away when the bees had worked out the comb, which would then hold the glass in position. This could, of course, be achieved with any design if the devices holding the four pieces of glass together were capable of being removed. Also 'Woodleigh', in his jottings wrote that he did not consider it just that those, including himself, who had made, used, exhibited and sold glass sections years previously, should have to procure a license from Hewitt in order to continue to do so. In his lengthy, but much edited response, Hewitt claimed the patent covered all glass sections "that are wholly or partly held in position by means of the honey comb within them". A license cost 2/6 and was required by all making them. At the time, he said that he was not going to sell them, but many years later he changed this saying it was his intention to sell them at 10/- per 100.

There followed a Cowan editorial indicating a change of direction, re glass sections, attempting to destroy their value – expensive, bees do not like them, and "anyway a manufacturer is about to market a different version shortly, so that will spoil the party". Prior to Hewitt's patent Cowan had described them as "very sweet, cleanly, and attractive". Hewitt's advertisements were declined by Cowan's magazines, effectively negating any possibility of financial gain from the patent. Hewitt later claimed that he only received 2/6 - one license!

This particular episode clearly demonstrates the problem that Hewitt was posing to the beekeeping establishment. It also illustrated Cowan's lack of ability in dealing with someone like Hewitt, who came from a totally different background. Most patents taken out for beekeeping equipment are not worth the trouble, beekeepers always find a means of circumventing them; and how could they be policed? Hewitt knew that he was not first with this – but he was first with the patent and he would argue that, unlike most of the B. B. K. A. Committee members, he needed to make money. Much later he would write that Cowan's actions made it difficult for him financially, something that says much about the true nature of Cowan, and the lengths to which he would go to if someone crossed him. For Hewitt it is never as clear-cut as he would have his audience believe. He later admitted that he was aware that there were others before him, quoting an American reference of 1864, conveniently an earlier date than Cowan claimed for his glass sections.

Following the bad press for Syrians, in 1885, Benton and Cowan turned their attention to Tunisian bees. The first attempt to transport queens failed, and it was not until 1886 that there was a successful importation by Hewitt even though he had claimed having them several years earlier. Fifty queens were sent, thirty arrived alive, some he sent to America for trial there, some he gave away in Britain, and some he considered unfit for use. He tried to sell a few at £5 5s each. In January 1887 Hewitt wrote in the *Journal of Horticulture* that Benton had sent him some Tunisian queens. The following month in the same magazine Hewitt renamed them 'Punic' bees, justifying this by pointing out that the bees came from Poenia (part of Tunisia was in that area), and the Romans had given the word 'Punic' to everything that came from there. They were apis nigra, literally 'black bee'. It is difficult to believe that Hewitt named these bees 'Punic' in order to deceive, otherwise why publicise the reason for doing it? There was a certain amount of logic in it.

Hewitt wrote profusely about Punics, extolling their many virtues as he saw them, and giving details on every aspect of their management. He was, of course, promoting them without having to pay for advertisements. He also gave queens to some well-known bee-keepers asking them to write, either to him, or in the magazines, reporting their experiences and opinions of them. In 1889 he sent four Punic queens to Broughton Carr, two died in transit, and of the other two Broughton Carr sent one to an un-named friend, but that died on her second journey. Broughton Carr wrote a very complimentary editorial in an issue of the *Record* in 1890 about the one he had received alive, and had successfully introduced into a colony.

Hewitt listed 32 qualities of Punic bees, many difficult to accept, for example No. 29, "If a Punic drone mates to any other race the resulting bees almost equal pure Punics, while it is a blend crossed the other way". Often the claims are somewhat contradictory; they swarm earlier than others but they are the least inclined to swarm. However, the most outstanding claims were that they did not suffer dysentery, and were immune to foul brood. This meant that there was no problem with them in Winter, and they removed the need for legislation to inspect and destroy diseased bees. Both these topics, Cowan considered his territory.

As well as the proliferation of articles on Punics, Hewitt began to advertise them for sale. However, by 1889, Cowan had had enough, and instructed Henderson, the then editor of the *B. B. J.*, to refuse to publish all further articles submitted by Hewitt. Henderson would not agree to this, believing that the magazine should be open to all reasonable material. Henderson died later that year, and the following year Cowan induced Broughton Carr to assign his monthly magazine the *Record,* to him, and move to London to work on both the *Record* and the *B. B. J.* Cowan now controlled the whole bee-keeping press in Britain, and combined with his position as Chairman of the B. B. K. A. Committee, controlled British bee-keeping. Carr was in his employ and Huckle (B. B. K. A. paid assistant secretary and business manager at the *B. B. J.*), also, effectively, his servant. Those who were not "Cowanites" would have no real place in bee-keeping's establishment, and those who had previously upset Cowan, were about to be dealt with. Friends of Cowan had for some time encouraged him to deal with Hewitt, a persistent critic, and now he was selling bees the claims for which, would negate Cowan's teachings on wintering as published in his B. B. K. A. paper, spoil his sales of cures for foul brood, question the need for bee disease legislation, challenge his claims for Ligurians and generally weaken his position in the bee-keeping world.

Probably by accident, Hewitt was published again in the. *B. B. J.* for May. His advertisements for both the *Record* and the *B. B. J.* were also carried, and orders for his virgin queens flooded in. However, Carr and Cowan soon refused to publish both his letters and his advertisements, yet again. The letters were declined because they were thinly veiled advertisements, and the reason given for not publishing the advert was because of some of its content. When Hewitt sent it back and asked for the objectionable material to be marked, he received a reply stating that they had nothing further to add to the original letter. Hewitt guaranteed money back on the 'Punic' queens if a purchaser was not fully satisfied. With such a guarantee, there could be no logical reason for refusing to carry his advertisements. The declined advertisement was later published by Hewitt and shows that rejection of it could only have been part of a vindictive plan to put him out of business.

PUNIC QUEENS,

THE BEES OF THE FUTURE.

PRICES.

TO BRITISH ISLES.		TO AMERICA.
Virgins...	... 2/6 each ...	... 1 dollar.
Fertile Untested	7/.. ,, ...	... 3 dollars 50 cents.
Pure Mated ...	... 20/- ,, ...	... 10 dollars.
Pure Selected ...	... 80/- ,, ...	... 40 dollars.

Cross-mated, when on hand, 5/-

N.B.—ONE PRICE ONLY ALL THE SEASON.

All guaranteed proof against Foul Brood.

Every Queen sent per mail prepaid and guaranteed against loss in transit, introduction mating (virgin) or winter dysentery. Virgins will be kept on hand to send on receipt of orders, each one of which will be at least six days old when sent off. Fertile Queens will be sent in rotation as ordered, so first come first served, and each order must be prepaid in whole or part; the balance, if any, can be sent when queen is ready to send. This condition is necessary as we cannot keep Fertile Queens long on hand, nor is the weather to depend on.

A speciality will be made of untested Fertile Queens at 7/- each, which will be mated where none but Pure Punic Drones exist for five miles round; if, however, a few bees are sent us, we will send word whether the bees are pure or not.

Note Well.—If these Punic Bees are not liked after trial, all money paid for the Queens will be returned in full. Also note that not one Queen reared by Mr. Hewitt has been yet reported as producing dysenteric bees, nor has any person condemned Punic Bees after trying them. We Challenge the world to disprove these FACTS.

Hewitt advertisement that was refused

Probably the truth of the Punic bees claimed ability to withstand foul brood, lay more in Hewitt's system of management than any natural immunity. A few weeks later, an article by "A Bee Breeder" claimed Hewitt was breeding from "worthless specimens". It is probable, because of later evidence, that the "bee breeder" was Samuel Simmins who was embarking on the latest of his enterprises, having had little success dealing in honey or hives. Hewitt believed Simmins new business to be partly funded by Cowan; the letter was almost certainly requested by the senior editor. Hewitt penned a response to this letter for publication in the *B. B. J.* Carr refused to publish it, citing the tone as being unacceptable. Despite Hewitt's protests his response was never published.

A short article and an advertisement were accepted for publication later the same year. This appears to indicate that there is no well-defined policy towards Hewitt's offerings, or lack of cohesion in the magazine offices.

A REVOLUTION !

NO more Weak or Dysenteric Hives. VIRGIN QUEENS, bred from Pure Selected, or Imported Stock, by 'A Hallamshire Bee-keeper.' Punic, Carniolan, Cyprian, or British, post free, 2s. 6d. each. Sent out six days old. Guaranteed against loss in Delivery, Introduction, Mating, or *WINTER DYSENTERY*. Send Orders quickly, and say when wanted ; Cash will do afterwards. For report of those sent last year, see *Record* for June. Season will close August 7th. Address J. HEWITT & Co.. CAMBRIDGE STREET, SHEFFIELD. 1347

Hewitt advertisement 1890

Hewitt was on the loose again; his next letter covered several matters and drew the following editorial footnote:

> Our correspondent has the unfortunate habit of differing from 'most folks' on most subjects, and the awkward part is that he generally assumes or implies that every one who differs from him is wrong, while he is right. We are quite willing to give full consideration to his views, but they are only *his* views after all, and he should in common fairness remember this when writing......

It had never stopped the editors. Hewitt came back restating his claim made in his previous piece, ending with "…as long as you will allow me to make such a claim I will freely allow you to disbelieve me." As one would anticipate there was a footnote:

> We do not think it necessary to add anything to the foot-
> note ….., preferring that our readers who are interested
> in the matter may read the letter and our remarks thereon,
> and form their own conclusions. Our correspondent must
> forgive us for saying that the comments &c., he makes on
> the merits of Punic bees – coming from himself – are really
> a matter for our advertisement pages and have been passed
> over in consequence.

Hewitt believed that the reason behind Cowan accepting his advertisements again was because of the guarantees that he offered with the bees. Cowan certainly thought this would produce an avalanche of claims and much correspondence to his bee magazines, and destroy Hewitt financially; neither materialised.

During this year, (1890), Hewitt claimed to have sold upwards of 2,000 virgin queens at 2/6 each. Later in the year, a Cowan supporter wrote to the *B. B. J.* asking about success of introducing the virgin queens, something that appeared to have been difficult for some bee-keepers. Again Hewitt penned a response, but the only one published was from another bee breeder who had failed to supply many queens; probably Simmins again. Later the same correspondent asked in a very cynical manner, how the Hallamshire queens had survived in, what had been a severe winter. Again Hewitt's response was not published, and his adverts for 1891 were declined for publication in both Cowan's journals. There were no other responses and Hewitt was banned once again. Undeterred, he turned his attention to America, and initially appeared to make sales, but following a further article by Cowan they ceased, and many asked for their money back

Whilst Punic bees had been banished from the *B. B. J.*, the same could not be said of the *Journal of Horticulture,* 'The Hallamshire Beekeeper' and 'The Lanarkshire Beekeeper', both wrote frequently about them, the former quoting opinions from named individuals. This clearly annoyed Cowan to such an extent that he used the devious tactic of writing a letter in the *B. B. J.* under the commonly used pseudonym 'Inquirer', questioning the entries in the *Journal of Horticulture.* The editors then answered the points they themselves had raised. They started:

> As a rule we do not like to import into our columns
> controversies originating in other journals, but as our
> correspondent asks for information for the benefit of our

```
readers......
```

They added that they were not aware of the experience of those individuals mentioned, and "none of our correspondents have reported on Punic bees". They only had to read the *Journal of Horticulture* to satisfy the first point, and the second is obviously inaccurate, but he continued to give much information of the bees of Africa! Then, "We know of no such race" i.e. Punics. Adding later:

```
We know nothing about the so-called Punic bees and can give
no information as to their value.  Possessing as we do one
of the largest libraries of bee literature in the kingdom,
it is strange that we have never found such a race alluded
to.  The word Punic means faithless, treacherous – neither
of which should be considered good qualifications for bees.
```

These three sentences illustrate how badly this had been thought out, and results from his refusal to address the real point whilst being pompous, cynical and arrogant. He finishes by criticising E. L. Pratt, an American "queen breeder and dealer", and advising readers:

```
...to wait for reports from experienced and well-known bee-
keepers.  We shall take care to give any reliable information
that may come to hand and be of value to our readers.
```

To use a 'Cowanism'; all my intelligent readers will be able to dissect the last sentence and agree with me as to it's limited value!

The next issue of the *B. B. J.* contained an attack on H. Alley and his magazine, *American Apiculturist* saying that it was full of little other than adverts for and articles in praise of Punics and golden Carniolans, the bees Alley bred. Again, this was an 'in house' generated contribution. There followed a letter in support of Punics and casting doubt on Cowan's meaning of *Punic* writing that the Romans used *punica fides* as a synonym for falsehood. As usual the sole reason for its publication was to enable the editor to respond:

```
...It is derived from the Latin punicus meaning of, or
pertaining to, the Carthagenians; deceitful, treacherous,
faithless.   Punica fides Punic faith, the faith of the
Carthagenians,   meaning  perfidiousness,   unfaithfulness,
treachery.   Punica fides was applied to the faith of the
Carthagenians because they believed in the perfidy of the
latter.
```

So all those derogatory terms must apply to a bee with that label? Cowan refused to accept Punic as a separate race of bee, but Hewitt did not claim such, he had simply given a bee from a particular part of Africa that name, and stated the reason. There had been no trouble calling the Italian bee, Ligurian.

In a further article that was widely published in American bee magazines, Cowan repeated that he believed there were no Punic bees in Tunisia, and people should not part with money until he had confirmed this. There were also claims that Americans had sent money to Hewitt but had not received queens or money back, something he strongly challenged, but no one would publish his response. It was a co-ordinated attack. Newman the editor of the *A. B. J.* and Cowan were very friendly at this point in time. Hewitt claimed that he was left with 5,000 queens unsold, as a result of Cowan's actions.

Other than his own leaflet, the only outlet open to Hewitt for his writings was the *Journal of Horticulture*, where he had been published on many occasions previously. Although Cowan and Dr Robert Hogg, the editor of the *Journal of Horticulture,* superficially, were on friendly terms, Cowan had wanted, and indeed expected, that Journal's bee column to be discontinued when he took over the *B. B. J.* Instead it became the voice of the disenfranchised (non Cowanite) beekeepers, edited at one time by Cheshire, Cowan's long-term enemy. The animosity that existed between Cheshire and Cowan at this time can be judged from part of a letter Cheshire wrote to another beekeeper, "he (Cowan) does not have the characteristics normally associated with an English gentleman and I studiously avoid him and his papers".

During the period in question the main correspondent and bee editor for the *Journal of Horticulture* was 'A Lanarkshire Beekeeper' – William Thompson, Auchinraith, Blantyre. When he died, in November 1898, his obituary in the *B. B. J.* was very short and included:

```
His views did not find much favour with English bee-keepers......
including us, ............. as a bee man he made his mark in the
history of the craft, and he will be much regretted.
```

Another of those printer's errors or a Freudian slip?

Cowan had written to Hogg asking for Thompson to be removed from the *Journal of Horticulture's* columns, claiming that he was not being truthful on many of the items he wrote about, and offering staff from the *British Bee Journal* as replacements. Hogg asked Thompson for verification of certain matters, and allowed him to continue. Hooker was persuaded to be the voice of the beekeeping establishment in the *Journal of Horticulture* and had entries most weeks, generally countering much of what Thompson and Hewitt wrote. Hogg had compromised.

In the *Journal of Horticulture* for the latter half of 1891, Thompson wrote extensively about his three Punic colonies, relating his generally favourable experiences. As usual Hooker responded. The intensity of the debate gathered momentum until C. Russell, of Ingmire Hall; a 16th century pile in Cumbria; wrote to the *B. B. J.* in response to an article by 'Expert'; (another 'in house' pseudonym), in which foreign bees were condemned, asking him what experience he had with the bees he condemned. In the response 'Expert' admitted that he had not kept Punics but "my experience as given, is borne out by most of our advanced bee-keepers......". leaving little doubt as to whom 'Expert' was, and despite there being no such correspondence published in his magazines. Russell responded but it was not published in the *B. B. J.* It was in the *Journal of Horticulture!* Some extracts:

```
Your correspondent "expert" entirely evades my question,
and I assume he is either unable or unwilling to give a
clear answer.  I read for information ...... this should come
from experienced persons.  Now, according to "Expert's"
own showing, what he teaches is second-hand.  ...... What does
your correspondent mean by "They are said also to gather a
great deal of propolis or anything sticky as a substitute".
As a substitute for what?  He is quite right, the Punics
are great breeders, and are as great honey gatherers, but
"Expert" is entirely wrong in saying they are tender; they
are very hardy bees.  ...... I would like to hear how "Expert"
can control swarming with the old black bees more than
with the the other varieties, and, more over I would like
to learn where the pure black bee is to be had in Britain.
...... Stick to your own experience , and give us reliable
information.
```

If that was not bad enough, editor Hogg added:

```
[We cannot understand why this letter is not published,
and are loth to believe it will not be published in the
contemporary to which it was addressed.  It is a very proper
letter under the circumstances, and was called for by the
one preceding it.  It is in our opinion the best letter of
the series, practical and precise, in which the writer has
obviously no motive but to elicit the truth.]
```

The report of the Doncaster Show, late June, 1891, in the *Journal of Horticulture* was written by Hooker. The majority was factual listing the awards, but towards the end he wrote:

```
Mr. J. Hewett, of Sheffield exhibited metal frame ends and
frames and a rack with glass sections, also a frame of his
```

Punic bees. If these Punic bees are exhibited again it
is to be hoped they will have at least three frames in a
suitable hive with bees and their queen, having brood and
sealed honey, properly ventilated, with the means of egress
and ingress similar to those observatory hives to which
prizes were awarded.

Hewitt had exhibited Punic bees but not in competition. His other exhibits were not prizewinners, but it was not usual to highlight a loser in such a report. Cowan was Steward at the show, W. Broughton Carr was a Judge.

Hooker was not the best choice to deal with Hewitt and Thompson, often getting the detail wrong enabling them to respond with cynicism. For example he wrote that there were "a good many meetings" of the committee to decide upon the standard frame; in fact there were two, and he was a member! He also got the judges at the Caledonian Apiarian's Society show wrong 'conveniently' missing out Abbott. He lacked the touch, and possibly arrogance, of his master.

In the September 3rd 1891, issue of the *Journal of Horticulture,* Hewitt responded to the article in the *B. B. J.* where Cowan had again answered "Inquirer", with a piece headed *Punic bees and those who know nothing of them.* Hewitt pointed out that: -

-   W. B. Carr in the *Record* for June 1890 wrote that his Punic stock was the "best and strongest" in his apiary.
-   Punic bees were mentioned in the *B. B. J.* on June 5th 1890.
-   When J. W. Woodley wrote in *B. B. J.* of 30th October 1890, condemning Punics before he had seen any, he (Hewitt) had written to the magazine offering to supply him with two queens to be tested against any other two he chose. It was declined for publication, as was the offer.
-   Several people other than E. L. Pratt of America and himself had spoken in favour of Punics, naming Henry Alley as at least one other.

Cowan contacted Hogg, claiming that the final paragraph of Hewitt's letter contained a libel. The claimed libel was obviously not in the final paragraph of the letter, illustrating inadequate care in checking the facts before acting upon them.

With the fuse lit, not everyone retreated the necessary distance.

18th Sept. Hogg met with Cowan to discuss the matter.
19th Sept. Cowan writes to Hogg, effectively an 'aide memoir'.
22nd Sept. Cowan to Hogg again, hustling.

23rd Sept. Hogg to Cowan enclosing copies of the article, requesting he mark the libellous parts. He repeated this request a few days later.

By early October, articles in the *Journal of Horticulture* from Hewitt no longer appeared, but Thompson continued in much the same vein.

11th Nov. Cowan's solicitor Ellis to Hogg. Ellis was writing on behalf of Cowan and Carr. The libellous charges against Cowan and Carr are obvious, and the insertion by Hogg must have been due to a slip by him. Withdraw publicly or litigation follows. Cowan & Carr will accept publication of apology in form they approve in *B. B. J., Journal of Horticulture, Bee-keepers Record* and one other. Hogg to pay £10 10s legal costs.

12th Nov. Hogg to Ellis. Grovelling. He had clearly believed all that Cowan had said.

13th Nov. Hogg to Ellis. Points out that he had written to Cowan twice asking for the relevant paragraph to be pointed out. He added that he was now aware of the controversy between Cowan and his correspondent, and encloses the apology that he, (Hogg), proposes to publish. It is not clear how Hogg discovered the problems that existed between Hewitt and Cowan, he certainly did not wire Hewitt to check the facts nor communicate with Wright, his deputy, who was in contact with Hewitt, and had been made aware of the facts of the matter as Hewitt saw them.

14th Nov. Ellis to Hogg. Letter received.

16th Nov. Ellis to Hogg. Apology too vague so not acceptable. One that they will accept enclosed. To be published also in *American Bee Journal, Apiculturist, British Bee Journal, Bee-keepers Record, Journal of Horticulture.* These were the publications in which the article or substance of it has appeared. All remaining copies of 3rd Sept. 1891 edition of *Journal of Horticulture,* to be destroyed. Once Hogg had capitulated the ante was upped somewhat, and what purpose would be served by destroying unsold copies of the magazine now six weeks old?

18th Nov. Ellis to Hogg. You have not replied.

18th Nov. Hogg to Ellis. Have just got back into the office. Next issue of *Journal of Horticulture* was already made up but it has been changed to the version you sent. He clearly felt obliged to respond to the pressure being applied.

18th Nov. Ellis to Hogg. Re-states the other conditions to be complied with and costs of £12 to be paid.

19th Nov. Apology appeared in the *Journal of Horticulture.* There was no heading, it read:

We have received notice that the article which appeared

on page 211 of the *Journal of Horticulture* for September 3rd, 1891, under the heading of "Punic Bees and Those Who Nothing About Them," and signed " A Hallamshire Bee-keeper," charges the editors of the British Bee Journal and Record, Messrs. Cowan and Carr, while purporting, in answer to an enquiry, to give all the information in their power about the so-called Punic bees, thus deliberately suppressed facts within their knowledge, and thus gave a false account of matters of interest to the readers of their journal. That the article also insinuates that Messrs. Cowan and Carr have some personal and unworthy motive for concealing facts which it is plainly stated they must have known. There was no mention of the Punic bees in the *Record* of June, 1890, nor has there been any allusion to them either editorially or by any of its correspondents. There is also no statement in the *Record* for June 1890, that Mr Carr, had a Punic stock in his possession, and he has never written anything about Punic bees.

It is suggested that the appearance of the article in question might be due to an oversight, and not to any intention to injure anyone. That is certainly the case, for it is far from our desire to make reflections on the reputation of those for whom we have never entertained feelings other than those of true respect, and we now desire to express our regret that the article referred to appeared, and to withdraw all the charges and insinuations therein contained.

It was printed in the *B. B. J.* on 26th November, under the heading "An Apology", with the following introductory paragraph:

We have been requested to insert the following apology from the Editor of the *Journal of Horticulture,* which appears in the issue of that journal for Nov. 19th inst., p 442.

I am not aware how many of the other conditions were complied with, if any. Newman published the apology in the *American Bee Journal* on the 10th December, and incorrectly, stated that all the unsold copies of the *Journal of Horticulture* had been destroyed, indicating that Cowan had informed him of one part of the saga but not the follow up.

By publishing the apology Hogg was in effect, saying his correspondent Hewitt, was a liar. However, Hewitt had already produced the evidence to prove his case to Wright, and this was communicated to Hogg the very next morning. Whilst questions arise as to the quality of the communications within the *Journal of Horticulture* office, much of the problem lay with Cowan's hustling for the apology to be published, and other than believing in some form of conspiracy, it is difficult to understand the desire for speed.

Whilst Hewitt wrote in an unsophisticated manner he generally gave accurate information and quoted the source. His derivations from those facts might be considered illogical or bizarre, but the basis was generally correct.

In the December 3rd, 1891, issue of the *Journal of Horticulture* the bee-keeping column was headed by the following:

AN EXPLANATION

It has come to our knowledge that the publication of a statement which appeared on page 442 in our issue of November 19th was inserted under a misapprehension, and that the information supplied to us 'there ......correspondents', is not in accordance with the facts. Here are the facts: -

On page 74 of the *Bee-keepers Record* for June, 1890. a question is asked by Guillaume, Wigtownshire, N. B., 'What kind of bee is the Punic?' to which the editorial reply is, 'According to a Hallamshire Bee-keeper the Punic bee comes from North Africa. It is dark in colour, and from our limited experience of it is a good worker and a prolific sort.'

It will be seen that 'Punic bees' are twice mentioned there, and that a reply is given founded on 'experience'. Yet in the *British Bee Journal* for August 27th, 1891, on page 381, the same Editors, in reply to 'Inquirer,' say, 'We know nothing about the so-called Punic bees, and can give no information as to their value.'

We were induced to publish the statement last week in reliance on the accuracy of the matter furnished, and with a desire to be just to the conductors of our contemporaries; but we must also be just to our correspondent 'A Hallamshire Bee-keeper,' who has placed in our hands evidence which justifies us in making this explanation; and we would add, our long experience has led us to regard him as an accurate correspondent.

This, of course, implied that Hogg considered the remarks, that Cowan claimed libellous, were true. Cowan considered that the publication of the explanation by Hogg effectively repeated the libel. There is no information available on whether Hogg gave Cowan notice of publication of this statement.

Cowan met with Hogg applying pressure in an attempt to get him to withdraw the explanation, taking great exception to the last paragraph, which effectively meant that Hogg believed all the criticisms of Cowan by Hewitt, not just those in the offending article. But both Hewitt and Cowan could not be telling the truth. Hogg challenged Cowan to produce evidence of Hewitt's inaccuracies published in the *Journal of Horticulture*. Cowan and Carr's attempt at this was pathetic. The saga continued:

7th Dec. Cowan to Hogg. Encloses his "explanation" in response to Hogg's "explanation" and demands that it be printed in the *Journal of Horticulture,* stating that he intended inserting it in his magazines. Cowan also states that Carr conducts the *Record* and he is solely responsible for the *British Bee Journal.* This appears to have been an attempt by Cowan to clarify something that had occurred in his discussions with Hogg, or more sinisterly Cowan distancing himself from the remarks that appeared in the *Record.*

8th Dec. Cowan to Hogg. States that he had not consulted Carr on this, - he had now! Carr was distancing himself from the quarrel, and was not going to be Cowan's 'scapegoat'.

9th Dec. Hogg to Cowan. You cannot dissociate yourself from what appears in the *Record.* It is beginning to look like Cowan has met his match. Hogg offered to publish Cowan's 'explanation' in the *Journal of Horticulture* when Cowan printed his in the *British Bee Journal.* He suggests that logically Hogg's should appear first followed by Cowan's as a rejoinder. Given the circumstances existing at that time, this was a generous offer.

12th Dec. Cowan to Hogg. Insert my 'explanation' and include a satisfactory explanation of the last paragraph in your 'explanation', or I will take steps as advised and issue a 'writ', presumably for libel. If Cowan had, as he states, taken legal advice prior to this it is difficult to understand why his solicitor had not conducted this correspondence. Therefore, the conclusion must be that this is Cowan's view of the law, or his solicitor was not prepared to write the letter.

16th Dec. Hogg's solicitors, Webb, Burt and Nicholls to Cowan. Repeats Hoggs 'reasonable' offer.

18th Dec. Ellis to Webb. He has met with Cowan and suggested to him that Webb's suggestion be acted upon. A printer's proof of what Cowan is to insert will be forwarded. The question here is, why was this necessary? Hogg and Cowan were both in possession of the two 'Explanations'.

19th Dec. Ellis to Hogg. Encloses printers proof of article as it will appear in the *British Bee Journal.* and *Record* next Thursday, and asking that the explanation which Cowan sent Hogg on 7th Dec. appear in the *Journal of Horticulture.* Also, Hogg should explain and withdraw insinuation in the last paragraph of the explanation that appeared in the *Journal of Horticulture.* If he withdrew it why was an explanation required? It was a request that was never going to be acceded to, there being no logical reason. This is the behaviour of a very arrogant individual. Cowan, at this junction was really on the defensive, and should have let the matter drop long before this point. It is also clear that Cowan was not taking the advice of his solicitor.

23rd Dec. Ellis and Webb had meeting.

23rd Dec. Webb to Ellis. Your letter of 19th did not arrive until 21st. We cannot permit explanation as sent and cannot agree to add the explanation of last paragraph. We have amended printer's proof to make it acceptable. It will appear in the *Journal of Horticulture* after it appears in the *British Bee Journal.* If it is published in the form sent, Hogg will not publish it at all.

23rd Dec. Ellis to Webb. Too late, it has been published, but Cowan was fully justified and you should publish it also.

24th Dec. Webb to Ellis. Printers proof should have been approved prior to publication. Hogg to be consulted after Christmas.

The content of the printers proof appeared in The *British Bee Journal* of 24th December, 1891. The opening paragraph read:

> Our readers will have seen an apology on page 533 of the *British Bee Journal,* which was inserted at the request of the editor of the *Journal of Horticulture,* with reference to a <u>libellous article</u> which appeared on page 211 of that Journal. We regret that through an unintentional oversight (is there ever an intentional oversight?), on our part, an error has <u>crept into</u> that apology, and it is with pleasure that we withdraw any imputations or insinuations that may be conveyed in this particular paragraph. We were surprised that, without our having an opportunity given us of correcting the mistake, the following appeared on page 485 of the *Journal of Horticulture:* -

Although there is no published record, it is unlikely that Hogg would have accepted the underlined parts. The argument was about whether the article was libellous. The error did not 'creep in'; it was consciously inserted by Cowan.

The 'explanation' by Hogg came next, as quoted earlier.

There then followed 'An Explanation' from Cowan thus:

To the Editor of the 'Journal of Horticulture'.

I have had the *Journal of Horticulture* sent me, and my notice has been directed to an editorial explanation on page 485.

In your zeal to do justice to your correspondent you certainly appear to embrace the opportunity of being unjust to your contemporaries. When I wrote to you on the 19th September last 'I can find no allusion made in the Record about Punic bees, either editorially or by any of its contributors,' I did so in perfect good faith, but I frankly admit that I was led into the error by an unintentional oversight on my part. To have admitted that Punic bees had been once alluded to editorially in the Record would not have altered the nature of the libel[2], as there was no question about this paragraph, but about a reference by your correspondent to an article in which there was no mention of Punic bees at all. Any unprejudiced person can see that the editorial allusion you quote has nothing whatever to do with personal experience, as it is couched in general terms, just as the same term 'our limited experience' is daily used in a legitimate way in ordinary language in connection with things the writers have had no experience about, but about which they are writing from the reports of others. Such language is in daily use in all our papers, and it would be a stretch of the imagination to attribute personal experience to the writers.

The context will also show that it is more the expression of the limited experience of 'A Hallamshire Bee-keeper,' who had just been given as the authority, as he was the only one who had said anything about Punic bees, and this in the concurrent number of the *British Bee Journal*. This reply given by Mr. Carr could not have been from personal experience, as he had never seen a Punic stock, and knew nothing about such bees; in fact, although Mr Carr was anxious to do Mr Hewitt full justice, so little value did he attach to these bees that there is no reference to them in the index, hence the reason for my overlooking the paragraph referred to. Mr Carr seemed perfectly oblivious about this paragraph, and I am not surprised at it. The first description of Punic bees was given by Mr Hewitt in *B*.

---

2   "Libel" replaced by "statement"

B. J. on page 271,1890. He again refers to them on page 511, and on page 512 Mr J Luck asks for results respecting Punic bees from those who have tried them. There was not a single response to this appeal from any one who had tried them except for Mr Hewitt himself. He had already had full justice done him, and we naturally wanted to hear about the experience of some one else besides Mr Hewitt. No one will be surprised that after such a complete failure of reports, Mr Carr should have dismissed the matter from his mind as being of no importance.

Now, Sir, I think I have just cause for complaint. When I placed all the papers before you (including the *Record* of June containing the passage now referred to), you promised to have the matter investigated, and in your letter to me on 23rd of September you stated, in reply to my letter of the 19th of September: 'when Mr Wright returns to the office I will authorise him to see to the matter.' Mr Wright was to have returned about the 2nd of October, and not hearing from you, I wrote on the 22nd of October, and from your reply on the 23rd of October I gathered that you had not made the investigation which you had promised to make, although a month had elapsed. Had you made the investigation probably the error, which I am sorry was made, would have been detected and avoided. (So presumably Cowan's error was Hogg's fault!).

You refer to this mention of Punic bees and take the opportunity of saying: 'In the *British Bee Journal* for August 27th, 1891. page 381, the same Editors, in reply to "Inquirer", say, "We know nothing about the so-called Punic bees, and can give no information as to their value."' You Mr editor, are perfectly well aware that this is not true, and that[3] the reply to 'Inquirer' was written by me alone, and that I was speaking of my own experience, and not Mr Carr's. You will see by the *British Bee Journal* that I am responsible for the conduct of this paper, while Mr Carr conducts the Record, so that there was no inconsistency in my reply of not knowing anything about so-called Punic bees, and you must disconnect Mr. Carr, the writer of the reply in the Record, from my reply in the British Bee Journal altogether. But, after all, one of the main points

---

3   "You Mr Editor are perfectly well aware that this is not true and that" replaced by "This"

at issue was the statement in the article by Mr. Hewitt in the *Journal of Horticulture* on the 3rd of September in which he says: 'They do not say that the Punic stock in Mr. Carr's apiary in the spring of 1890 was the "best and strongest" he had' (see Record for June, 1890). On reference to the article from which the quotation is made by Mr. Hewitt, it will be seen that there is no mention whatever about Punic bees or to a Punic stock in Mr. Carr's apiary. The article refers to quite another matter altogether,[4] namely, to virgin queens sent out to test the wintering theory of the writer's. I have simply to emphasise what I have already said, that I know nothing about so-called Punic bees, although I know North African bees very well, more especially those of Algeria, Morocco, and Tunis. Some of these days, I shall have plenty to say about the bees of this last place, also about the apiary belonging to a French gentleman (whose name, for obvious reasons, I withhold), and who exports Tunisian bees, and whose apiary I intend visiting during my rambles in Africa, whither I contemplate going for the purpose of finding out why the bees of Tunis are not pure like those of the same varieties in Algiers, and why they sometimes show yellow bands.

[5]Your correspondent's accuracy is certainly very much shaken when we remember what he has said about Syrian and Cyprian bees, and we all know how they turned out. Also, for instance, what he says on page 316 of the *Journal of Horticulture* about these (Punic) bees 'that their natural months for rest is our summer, which is their winter.' I should think you, Mr. Editor, at any rate, would know that the seasons in North Africa, as in all places north of the equator, are at the same period of the year as ours, yet endorse this as accurate. – T. W. Cowan, Ed. British Bee Journal. – 7th Dec 1891.

By arrangement with Dr. Hogg,[6] the above explanation will be printed in the *Journal of Horticulture*.[7]

I have reproduced the complete letter so that any reader, so inclined, is able to dissect it, and generate their own opinion of its contents.

---

4   Dash inserted
5   Complete paragraph omitted
6   "Dr Hogg" replaced by "The Editor"
7   Sentence inserted above the signature

29th Dec. Webb to Ellis. Hogg will not publish the explanation as per printers proof and published in the *British Bee Journal*, and intends in next *Journal of Horticulture* to state that your publication of the explanation was premature. Hogg is entitled to an apology for what has been published, - absolutely no chance!

29th Dec. Ellis to Webb. Cowan only insisted that the "explanation" he sent on 7th Dec. be inserted, and not the one he published in the *B. B. J.* An extraordinary statement - what logical purpose could be served by having the two different versions of the same thing?

31st Dec. *Journal of Horticulture* publishes as threatened opening with:

> We have to state that the article which appeared on page 582 of the *British Bee Journal* of the 24th inst., headed "An Explanation," and signed T. W. Cowan, Editor *British Bee Journal*, was premature. We had consented to allow Mr Cowan to publish an explanation in our columns, and a proof print was sent to us for approval. This proof print was returned by us on the 21st inst. With certain alterations, some sentences being expunged which we considered libellous, and the actual form in which we finally agreed to publish it was as follows:

The footnotes detail the difference between the version published by Cowan and that by Hogg.

Whilst it is possible to compare the two explanations above, the same is not the case for the version that Cowan sent to Hogg on the 7th December, but which Hogg claimed to be different again.

4th Jan., 1892. Ellis to Webb. Cowan is to write a protest to Hogg. This appears to show that Cowan was displeased or frustrated with Ellis, his solicitor, and probably not accepting his advice. The protest is as would be expected, reiteration of the facts as Cowan saw them, still claiming that the contents of the initial article that gave rise to all this was libellous.

8th Jan. Webb to Ellis. (i). Hogg's offer to Cowan had not been accepted. (ii). Printers proof was different than the original sent by Cowan. [If the printer's proof was not sent for approval, why send it at all? This does not reflect well on Cowan, and it was the second time that he had resorted to this tactic! Arrogance]. (iii). The additional explanation Hogg refuses to give. [This shows a complete lack of understanding that Hogg was in no position to give this. More Arrogance.]

14th Jan. Ellis to Webb. This is repetition – again! However, Cowan now claims that the change between the first and second 'explanations' expressed

Cowan's regret for the unintentional error that had crept into the *Record* page 172.

[So why was this not stated at the time of sending? The irony here is that this was the proposition put to Cowan by Hogg many letters ago as a means of withdrawing with grace, but declined.

The basis for this was that Cowan (and Carr?), had only checked the index of the *Record* for 1891, and made the statement that Punics had not been mentioned in that publication. Carr would have been responsible for maintaining the index for that magazine, apparently ongoing, so Cowan once again appears to be placing the blame with him. It is an excuse that was unlikely to convince many; after all, how long would it take to check the actual contents of one month of the *Record?*]

Ellis added that Cowan had not insisted that Hogg make a further explanation of the last paragraph of his 'explanation'. Oh yes he had!

14th Jan. *Journal of Horticulture* published an article headed "Mr. T. W. Cowan's Inaccuracies". It read:

We have been asked by Mr Thomas William Cowan to correct some inaccuracies which have appeared in this Journal, and in view of certain eventualities Mr Cowan has also intimated his intention to publish certain correspondence in his papers. It is immaterial to us what he publishes, so long as he refrains from libellous statements. There was a misprint on the page mentioned below of 21st instead of 23rd, but we select as by far the most important of the inaccuracies the following, which were supplied to us for publication by Mr Cowan, and we accepted them in the belief that hey were correct.

In the statement published by us on November 19th there were four facts asserted, viz. 1st, that Punic bees had never been alluded to in the *Record* by any correspondent; 2nd, or editorially; 3rd, Mr Carr had never written about Punic bees; and 4th, Mr Carr never had a stock of Punic bees.

These are the statements alluded to above as having been supplied by Mr. Cowan, who insisted on their insertion in our columns. The first three he now admits to have been incorrect in his "explanation" on page 571 of our issue of 31st December 1891; and as regards the fourth, we may say for the information of Mr. Cowan and our readers that we have evidence that proves beyond a doubt to our mind that

> Mr. W. B. Carr did have a stock of bees from a Punic queen
> in 1889-1890. We have asked "A. H. B. K." to allow us to
> publish some evidence that we believe he can supply, but he
> informs us that he is taking action against Messrs. Cowan
> and Carr for libel. And that it would be best to produce
> the evidence in Court, along with other evidence.

21st Jan. 1892. As a supplement to the *British Bee Journal*, Cowan published "the whole correspondence" on this matter "in order that our readers can make their own minds about it". It was, as would be expected of him, only part of the correspondence; there were significant omissions. At the end of it Cowan writes:

> Since the last of the above letters was written, our
> attention has been called to an article on page 31, of the
> *Journal of Horticulture* for January 14th 1892, headed 'Mr
> Cowan's Inaccuracies' (sic). We commend that article to
> our readers and would ask them to note its tone and very
> evident bias, - LITERA SCRIPTA MANET."

That Latin again, deliberately antagonistic – written letters endure. Obviously fighting to the end! He did not reprint the article from the Journal of Horticulture, in order that his readers could 'judge for themselves', hoping that his subscribers did not read the opposition magazine. He also omitted to include the fact that he paid damages and all legal costs.

If only duels had still been allowed in law.

I think Cowan must have thought that the matter would now have rested, but the *Journal of Horticulture* continued to publish Hewitt, so Cowan took the extraordinary step, attempting legally to stop his effusions appearing. He dragged out this action for some considerable time, during which Hewitt was silenced, but eventually it had to end. Cowan paid notional damages and all costs for this also.

Hewitt was now free to repeat the allegations without fear of any legal recrimination. However, Hogg refused to publish Hewitt's rants, limiting his material in the *Journal of Horticulture* to bee matters and specific responses, including all the evidence, following attacks on him in the *British Bee Journal*. This reduced the amount considerably. Hooker returned as the establishment's voice in the *Journal of Horticulture* locking horns yet again with Thompson and Hewitt. Normal hostilities had resumed!

Cowan wrecked Hewitt's queen trade for 1892 by again telling all his readers to refrain from buying Punics until he had gone to North Africa and was able to provide "reliable information" about them. He took this 'ramble' unusually with one of his daughters, it obviously not appealing to Mrs. Cowan. They visited many apiaries including those from which Benton alleged he was shipping queens. During the time Cowan was in North Africa, S. Simmins, who earlier had declared that Hewitt bred from "worthless specimens", advertised directly imported Punic queens at the same price that he was selling Syrians etc., highlighting the grossly inflated prices being charged by Hewitt. These bees do not appear to have ever come to market. It is apparent that they were fifty queens that Cowan attempted to send to Britain and failed, either because they perished on the journey, something that Hewitt/Benton had also experienced, or, ironically, the postal authorities refused to handle them, resulting in them being returned. Cowan never responded to the accusation that he was co-operating with Simmins in this failed venture, nor that he had invested in the same individual's queen business that became bankrupt in very quick time, returning very little to its' investors.

Upon his return Cowan wrote of his visit to North Africa, declaring, of course, that there was no such race as 'Punics'. Adding that 'Tunisians' belonged to the race Apis Mellifica, the common English black bee, which it resembled in appearance. Hewitt was financially crippled before this, and Americans Pratt and Alley had also suffered severe blows to their businesses.

Whenever the legal wrangle with the *Journal of Horticulture* was mentioned Cowan took the opportunity of saying "…it is only a gardening paper" - forced through clenched teeth presumably.

In May 1893, Hewitt published "An Explanation", even the title being deliberately antagonistic; he printed it himself. This was a pamphlet that he distributed to those to whom he sold queens, and anyone else that would accept it at shows etc. He listed 29 points of fact as he saw them. There was repetition, but it was very factual, and in most instances sources were quoted, but as usual with Hewitt he starts in a perfectly rational manner, but develops it into somewhat bizarre conclusions. A prime example of this is his belief that Cowan's reason for wanting an act of Parliament to destroy bees with foul brood was to eliminate his Punic strain, which he alleged were immune to foul brood if managed as he recommended. He actually accused Cowan of deliberately spreading foul brood to bring about pressure on Parliament!

Whilst asking for all who received his leaflet to tell their friends Hewitt also requested that they apply for shares in a new Journal he was to publish. Presumably there was little interest in the Journal shares because it was not until May 1897 that vol. I, no. 1 of *The Bee Master* appeared. Again, it was, printed and mostly written by Hewitt. It appeared spasmodically for a total of seven editions, the final issue appearing in July 1898. Hewitt stood little chance with his magazine; Cowan threatened distributors with legal action if they handled it, and refused to take advertisements for his own publications from those who also placed them in the *Bee Master*. It never achieved a circulation greater than 300 per issue, although 10,000 of no. 6 were allegedly distributed. The only external contributions were from others who felt wronged by Cowan and Carr. The final issue limped out in July 1898.

The *Bee Master* contained little bee matter, comprising mostly of Hewitt and Thompson's attack on Cowan and his clique. There was much repetition and the continuing threat that there was more to come, but it never materialised; it was just an outlet for their anger. It was a full frontal attack on Cowan and Carr and those who supported them and it is difficult to believe that parts, especially towards the later issues, were not libellous. Cowan did not respond, he had been forced to retreat on two occasions and was not going to make the same mistake again. He was obviously of the opinion that he had dealt with Hewitt. He had certainly landed a near fatal blow to Hewitt's bee business. In what appears to be an attempt to bring closure to the Hewitt affair, in answer to Woodley (a devout Cowanite), who had inquired about advertisements from 'H' disappearing from his magazines, Cowan stated that he had "died some months ago". Although not specifically named, Hewitt deduced this had to refer to him because all the other H's that advertised were still very much alive and having their advertisements included, and why else would the full name not have been given? It was of course untrue; Hewitt was very much alive.

Banner of *The Bee Master*

In his book *In Search of the Best Strains of Bees* Brother Adam incorrectly states that Hewitt visited Tunis and brought the North African bee to the notice of English beekeepers under the name of the Punic bee, adding that in North Africa it is commonly known as the Arab bee adding:

Its main centre of distribution is undoubtedly in the high
ground known to the Arab as *tell*; the name *Tellian* …would
therefore seem to be the most appropriate." "…the *Tellian*
bee is of no value to the amateur beekeeper. But there
seems little doubt that it is one of the most valuable
races for cross-breeding.

He summed the Tellian bee as:

(1) extreme swarming tendancy, (2) a highly developed
susceptibility to brood diseases, common with most of
European dark bees. (3) a lavish use of propolis, (4) watery
cappings. Against these defects must be set unparalleled
stamina, fertility and foraging power………all the findings I
have made up to now indicate that the Tellian is a primary
race, and that the numerous varieties of brown or black
bees – at least those of Western Europe – have in the course
of time evolved from the Tellian.

Hewitt would have been pleased with all but (2). He had come to the
conclusion that selling virgin queens to be mated with the local variety was the
way forward.

There were other ingredients to the Cowan/Hewitt mutual hatred.

Live bees by post.
Cowan, representing the B. B. K. A., had petitioned the Post Master General, and
over a protracted period met with his representatives many times, in an attempt
to get bees, particularly queens and her escorts, accepted for transmission
through the normal post. He had failed on every occasion. The Post Office
regulation in this respect banned dangerous or offensive articles, unless they
were packed in such a way as to be secure and impossible to injure anyone.
The regulation also covered a package that enclosed any living creature, which
is either noxious or likely to injure either post or postal officials. The Postal
authorities also had sole right to carry packages below one ounce in weight.
Hewitt argued that a queen with a few attendants in a cage should not be
banned under these regulations, and sent many such packages. If the Post
Office suspected they were bees they sent the package to the Dead Letter
office, and sent a note to the addressee to collect it. At no stage did they
attempt a prosecution against Hewitt even though he had been sending bees
by letter post since 1883.

When it appeared that the B. B. K. A. were about to get what they wanted
from the Post Office, in February 1892, Cowan wrote to them asking why Hewitt

had been allowed to send live bees by post which was contrary to the law. For an organisation that was requesting just such a thing, this letter gave the impression to the Post Office that the B. B. K. A. objected to bees being allowed to be carried by the ordinary post. It was, of course, Cowan, trying to shut down Hewitt's bee business. Cowan had signed the letter "Thos. Wm. Cowan, Chairman of the British Beekeepers Association", which is inaccurate, but more seriously there had been no discussion within the Committee of the Association, resulting in the letter being sent. It was a private communication. Cowan was asked to a meeting with some officials of the Post Office the following month, at which he later claimed to have been informed that bees could be sent by parcel post. Amazingly he failed to inform either the B. B. K. A. or the *British Bee Journal* readers of this fact until 1897.

Coincidently, in early 1892, the Post Office warned Hewitt against sending bees by letter post, but informed him that it was legal to do so by parcel post.

The first two queens Hewitt sent by parcel post in 1894 were returned. Following his representations, the Post Office conceded that this was a mistake in his case, because he had been given prior authority to send bees by parcels. In April 1896 a Treasury warrant was issued that stated that "no living creature could be conveyed by the post except with the special permission of the Postmaster General".

In 1891 the International Postal Union passed a clause allowing queens in travelling cages to be allowed through the letter post. This meant that the British Post Office delivered any such package despatched from outside Britain.

During this whole saga, Hewitt had been communicating with his local M. P., Sir Howard Vincent, who in March 1897 raised the matter in Parliament, where it was treated with derision. A member of the B. B. K. A. committee who was also an M. P. was believed to have been the initiator of this. Hewitt immediately wrote to his local paper, as usual a very long, detailed, occasionally confused offering. The letter highlighted the inconsistency that the Post Office officials had shown in this matter, and stating that it did not reflect well on those involved. Sir Howard Vincent put a second question on the notice paper of the House of Commons that resulted in the Secretary to the Post Master General agreeing, this time, to give it consideration. On the 22nd February 1898, the P. M. G. issued a warrant allowing bees to be carried by post, effective from 1st March. In the *British Bee Journal* for March 3rd 1898, the editor wrote "It is at length our privilege to make the fact known that bees may now be sent to any part of the United Kingdom by letter post", which seems to claim the success as theirs. Obviously there was no mention of the part played by Hewitt or Vincent.

Many writers consider this matter to be the reason for the conflict between

Hewitt and Cowan. It is true that Hewitt was furious at not being given credit for the part he played in this matter, or as he saw it, being singularly responsible for the change in the law that made it possible to send live bees by normal post. In his writings he returned on many occasions to the theme that no matter what an individual's position in society, he should be given credit for his ideas and work, and in this respect Cowan robbed him of his true position in the bee-keeping world. There is little doubt that he had succeeded where Cowan and the B. B. K. A. had failed. However, he believed that Cowan did not want beekeepers to be allowed to send live bees through the post. As justification for this opinion he cited the contents of Cowan's letter to the P. M. G. and the fact that he knew in 1892 that live bees could be sent by parcel post but did not inform bee-keepers until 1897.

As always when Hewitt is involved there is intrigue, threats, and the setting of traps for his rivals. Hewitt had certainly seen the letter Cowan wrote to the P. M. G. in 1892, but how? It was not the only time he obtained documentation that was beneficial to his fight.

Buoyed by his success he concluded the matter thus:

> So far, *Bee Master* has done more for bee keepers than the *British Bee Journal* or all the Beekeepers Associations have done in the past. It has not killed the intended Bee Pest Bill, but should any person be ill advised enough to introduce it into Parliament, both he and the promoters will get such an advertisement as will ensure its prompt withdrawal for ever.

Foul Brood.

Hewitt believed that the main reason Cowan fought for bee disease legislation was that it would provide kudos and money for the B. B. K. A.s' first class experts, and thereby strengthening Cowan's position as head of the certifying organisation and chief examiner of the experts. Hewitt attacked on two fronts, highlighting the lack of foul brood expertise by the experts, and accusing Cowan of deliberately spreading foul brood by publishing adverts for bees known to be infected with the disease.

C. N. White was a schoolmaster, first class expert of the B. B. K. A., and member of the B. B. K. A. Committee; he wrote under the pseudonym 'Ivo'. In the magazine *Farm, Field and Fireside* he wrote of the destructive nature of foul brood if not dealt with. Photographs of an apiary accompanied the article illustrating it before and after foul brood. It was identified as White's apiary because the before picture was the same as that which appeared earlier in one of the regular series 'Homes of the Honey Bee'. So here was a first class expert with an apiary riddled with foul brood - White was an easy target for Hewitt.

White was found to be travelling on a train with stocks of bees that a fellow passenger smelt had foul brood. He asked White why he was moving bees that had foul brood, he responded claiming that he was a first class expert and his bees did not have foul brood. It was alleged that the two alighted at the next station, bees inspected, confirmed to have foul brood and destroyed by fire on the platform. White challenged the story, but was inconsistent with his defence. There was considerable correspondence between White and Hewitt. Other individuals were unwittingly caught-up in the controversy which was often about petty detail. White claimed to have evidence that proved the story incorrect but he was only prepared to present it to the Committee of the B. B. K. A. This was never likely to convince Hewitt – "….same as Satan convincing the Devil".

But White was already in trouble because in his address at the Royal Show of 1894 in Cambridge, he attacked the B. B. K. A. for not doing enough to contain the spread of foul brood in Cambridgeshire. Obviously this did not resonate well with the assembled B. B. K. A. members, and even less well received were his remarks about the central body for not doing enough for cottagers and excluding them from membership of the Committee. Cowan responded:

> ...a charge not only unfounded but absurd on the face of it. We should like to know what cottager bee-keeper could, or would, afford the time and expense of attending the monthly meetings of the committee, which must of necessity be held in London? ……… the committee had continuously laboured in the interest of cottagers.

White responded with a letter to the *British Bee Journal,* duly published, but subject to the usual withering footnote from Cowan.

White continuing his struggle with Hewitt would serve no purpose; he would not get support; he had effectively cast himself adrift from the central establishment of beekeeping - the matter was never resolved. In 1913 at a conversazione, someone remarked that it was good to see White attending a B. B. K. A. function again, to which he responded that he was normally too busy to attend meetings. He died a short time later.

Whilst the White affair did not have the desired effect for Hewitt re foul brood, he was on firmer ground in the Samson case. The Rev. Gavin H. Samson, B. B. K. A. member and expert, advertised bees for sale in the *British Bee Journal,* stating that they were healthy. A Mr. Hamlin Harris, B. B. K. A. expert, inspected them twice with a view to purchase, but withdrew because he considered them to have foul brood, and he informed the *B. B. J.* of such. The bees were bought by another individual who subsequently sued and won compensation from Samson. Hewitt believed that Cowan and Carr were culpable because they did not inform their readers that the bees had foul brood even though they were informed of such, proving that they deliberately set out to spread the disease.

A few months after this court case, Rev. Gavin Hamilton Samson appeared in court again, this time for examination in bankruptcy. He blamed his bankruptcy on the action brought against him in the previous case, and stated that the Archbishop of Canterbury had "inhibited him from doing duty" because of his misconduct. He further acknowledged that his wife wrote his sermons and that he sold them to other clergymen, earning a sum between £30 and £40 per annum from such actions. He also accepted that he attended horse race meetings and lost money backing unsuccessful horses.

In the wake of the *Journal of Horticulture* episode, Cowan had written a short piece headed "Birds of a Feather", which was directed at Hewitt even though he was not named. Hewitt referred to this article many times - the Samson case was made for it. It is also interesting to compare this to Cowan's mention of Father Meehan in his dispute with Digges.

In 1896 the *British Bee Journal* carried an advertisement from the Rev. G. Samson, Southlands, New Romney, selling his bee farm – no longer a 'bird of a feather'?

Allegedly a photograph of one of Hewitt's apiaries.

W. Herrod-Hempsall in *Bee-keeping New and Old* published a photograph of what he claimed was one of Hewitt's apiaries with the comment:

The grave menace of such an apiary to the bee colonies in
those properly managed situated near by is so obvious that
it need not be stressed further. (*Spero meliara*).

His mentor would have been very proud. However, he does not make it clear how the reader is to determine from the photograph that the bees had foul brood.

The Kirby Affair.

Hewitt sent samples of his Punic bees to the British Museum where they were displayed with the name he gave them attached. W. F. Kirby, an employee of the Museum, showed the display and spoke of the bees at the November 1891, meeting of the Entomological Society. This was reported in the *British Bee Journal* of 17th December, but Cowan carefully avoided quoting directly from the Transactions of the Society, of which he was a fellow, writing as if he had been present at the meeting, and seen the "poor dried specimens" declaring "..so far as we can judge they were no different from ordinary Tunisian bees - not a new species at all". After Cowan saw Hewitt's exhibition of a frame of live Punic bees at the Doncaster show of 1891, he wrote his Journal "..we could not see that they differed in any way from the ordinary North American bees", and on another occasion that they were the same as the British Black bee. Neither Cowan nor Carr attended the November meeting of the Entomological Society.

The remainder of the article was the usual Cowan offering, but he also mentioned the Kassatyr apiary of Tunisia run by a Frenchman. This detail had only been passed to Kirby in the letter that had accompanied the samples. Kirby confirmed to Hewitt that he did not show the letter to Cowan or pass on the information therein, leading Hewitt to accuse Cowan of somehow accessing the letter when he visited Kirby at the British Museum to see the bees.

Kirby insisted that Cowan publish a letter from him saying that in view of the contents of his entry in the *B. B. J.* on the Entomological Society meeting, Cowan must have misunderstood what he said. Kirby clearly thought that Cowan had been present at the meeting, adding that the Transactions accurately reported what he had said. Cowan appended the, by now, familiar insulting footnote – 'could not be interpreted otherwise, etc'. There followed anonymous contributions questioning Kirby's position at the British Museum, attempting to discredit him, after which he wrote to Cowan and Hewitt to say that he was not getting involved further.

Yet another completely unsatisfactory incident, illustrating that there was no limit to what Cowan would resort to in order to justify his opinion and retain his position.

Hewitt and the Baroness.

In February, 1892, Hewitt wrote to the Baroness Burdett-Coutts asking her to investigate certain charges that he levelled at Cowan. The only response was an acknowledgement. Hewitt believed that she spoke to Cowan about the points he raised, but it was doubtful if she ever questioned Cowan's version, because they had very great respect for each other and thought along similar lines. It is difficult to understand why she did not respond, in order, however remote, to attempt to draw a line under the affair. In 1897 Hewitt again wrote to the President in a similar vein, with the same result. I have not seen either of these letters, but they were probably of the same tenor as the open letter that appeared in the penultimate *Bee Master* in May 1898. Hewitt wrote, in what he would describe as an accurate down to earth style; not the 'norm' in Cowan's world and even less so in that of the Baroness.

The letter is reproduced from *The Beemaster* in Appendix 4. It is certainly down to earth in style, but not all of it is accurate. It is really a challenge: an attempt to get either the Baroness or Cowan to take legal action against him. They resisted the temptation.

If the above is viewed in conjunction with Cowan's known movements over this period it is not difficult to generate numerous conspiracy theories. From the time line and the dates above: -

February 1892, Hewitt writes to Burdett-Coutts.

May 1893, Hewitt publishes "An Explanation".

November 1896, Mr and Mrs Cowan travel to America, vaguely stated as for a 'long sojourn'. There is no evidence that any of the four youngest children travelled with them; Alexander was already in America.

May 1897, Bee Master launched.

September 1897, Hewitt writes to Burdett-Coutts again.

May 1898, Hewitt publishes open letter to Burdett-Coutts.

June 1898, Mr and Mrs Cowan return to Britain to arrange for family to travel to America and organise withdrawal from London address.

July 1898, last edition of *Bee Master.*

Mid October 1898, Helena and Herbert die en route to America.

Late October 1898, Mr and Mrs Cowan travel to America with Edith.

Early 1899, Mr and Mrs Cowan and Edith arrive Pacific Grove.

It is very tempting to speculate that Hewitt had made life for Cowan so uncomfortable that he decided to emigrate, taking his family with him, having scouted for the best location on the first trip. Hewitt taunted him saying that he was afraid to return to Britain, because he would have to 'deal with' him or 'loose face' in front of his cronies. On another occasion Hewitt wrote that Cowan had said that he was returning to Britain to 'finally crush' him. Cowan always gave the health of his wife or himself as the reason for most of his travels abroad. The truth as to why he went to America, apparently permanently, will, like most things never be known, but it was a strange choice given his regular lengthy periods in Switzerland, and following his previous harsh criticisms of America and its residents. Climate, relative value of money, a society that welcomed and revered him and a locality that suited, not just his health, but also catered for his interests, probably had most to do with the decision. Perhaps the Cowans were just enjoying their good fortune in life.

Hewitt did not need to be further crushed, he admitted to having been financially ruined by Cowan's selfish actions, but he did not go away. He continued to write for the *Journal of Horticulture,* but they only published his benign material.

However, in 1919, when Cowan and the B. B. K. A. were yet again trying to get Parliament to introduce bee disease legislation, the *British Bee Journal* called for all those who opposed legislation to state their views "free from all acerbity, personalities and recriminations". The belief by the bee-keeping establishment was that those who previously opposed legislation would by now, have changed their minds and communicate such. The editorial calling for opinions forcibly made the point that the Bill was drawn up by the Board of Agriculture, not the Association or any individual within it. This was recognition that it had been a focus for much of the earlier opposition. For those who had been involved from the very first attempt, when Cowan drew up the Bill whilst on his summer holidays, there was little noticeable difference.

Hewitt viewed this as his opportunity to get published in Cowan's magazines again. He, of course, opposed the legislation, and his first letter is typically Hewitt, certainly not complying with the rules of engagement as laid down. He relates the detail of the passage of the previous Bill in 1912/3, which does not agree with that given in the editorial quoted. In the penultimate paragraph he writes:

> I cannot conceive of any elected Parliament in this country ever passing a Bill to trample on people who cultivate bees as a business and successfully, too, on the clamour of a lot of ignorant upstarts, who cannot see any faults in the books of Mr. T. W. Cowan, Mr W. Herrod-Hempsall, or the No. 48 Food Production Leaflet. When the *B. B. J.* and the B. B. K. A. acts up to its professions and straightens out their past sins, I shall be then tempted to say what I know to be facts to the benefit of all. JOHN HEWITT, Bee Master, Sheffield.

As would be expected the footnote was longer than the letter, the only reason for printing Hewitt's 'effusion' in the first place. I can only assume that he sent the letter because he assumed that, as usual, it would not be printed, giving him reason for an entry in newly launched *Beecraft* or *Bee World*.

Cowan's footnote to Hewitt's letter contained the usual insults – "…imaginary personal grievances …vent their spleen…similar irrelevant matter….needless to say, our correspondent's assertions are untruthful", deliberately or otherwise, continued by misunderstanding the content of the letter, and responding accordingly, even interpreting the term "ignorant upstarts" to apply to cottagers!

Hewitt wrote again but it was not published, the editor imposed a limit of one letter per correspondent, which caused Rev. Tickner Edwardes, in one of his early letters to the *B. B. J.* to remark that he noticed Hewitt had been "hewn". However, with every letter having an editorial footnote attached, the rule clearly did not apply to the editors.

Hewitt's final contribution to the bee press appears to have been in the August 1922, *Bee World*, where in response to a previous editorial, he wrote "The Prestige of Beekeeping". This was a protest at individuals not getting their due rewards for their inventions and ideas in bee-keeping. In particular he highlights his claim to be the first to recommend feeding candy in Winter, which Cowan et al canvassed against. It contained the usual 'Hewittism' by claiming that he has never publicised the full recipe for the candy. So by denying him his rights bee-keepers have never reaped the full benefit. As was usual, it draws an editorial footnote, this time suggesting that he gets on with his life. But Hewitt had many friends and *Bee World* published several complimentary pieces about him.

In 1923 he was still advertising; *Bee World* carried this entry:

```
Punic bees - If you want healthy bees and reap good crops of
honey send a postcard to John Hewitt and Co., Bath Street,
Sheffield for particulars of these wonderful bees.
```

By now he was seventy-three years old.

Chapter VII

Bee Disease Legislation

In the latter part of the 19th century, diseases of the brood became a major problem for bee-keepers in most parts of the world. However, 'foul brood' probably existed in Britain in the Middle Ages. Certainly Remnant in 1639 was aware of it. In *A Discourse or Historie of Bees* he wrote:

> You shall know when there is dead brood or other putrefaction, by lifting them a little up on one side, when it is darke, and put your nose toward the worke, and so shall you find an ill savour:

Then after describing how to smoke them over an open fire he adds:

> Some putrification sticks fast in the cells or holes, and is of a brown or deep yellowish colour, which should have been brood, but came not to perfection.

Remnant also described chilled brood so there is little doubt that this was foul brood. His 'solution' was to cut it out. Classical writers had also encountered disease of the brood. Scientific knowledge was well short of diagnosing it, and the method of keeping bees until the advent of the moveable frame hive, did not lend itself to treatment, even if one had been known.

T. W. Cowan's tenure at the head of British beekeeping was at a time when bee diseases were a major challenge for both scientific and practical aspects of keeping bees domestically, resulting in considerable on-going modifications to bee management techniques. Ultimately, the advancement of scientific knowledge with the passage of time was likely to make someone who wrote on the subject look very silly. Cowan dispensed positive instruction on this, as with all issues on which he wrote, never tempering the advice with the proviso, that it was the best available at the time. His methods and the products he sold always 'cured' the disease. With the benefit of present day scientific knowledge, we know that the claims made, were very exaggerated.

I encourage readers to consult an expert detailed account of these diseases elsewhere. However, a brief resume of present knowledge will help put the efforts of Cowan's period into perspective. Not claiming expertise in such matters it has been checked for me by a Bee Disease Inspector who modestly wished not to be credited.

American Foul Brood.

American Foul Brood is highly contagious, easily spread, and will ultimately destroy the infected colony. The symptoms are a change in larvae colour from white to light brown then dark brown, and finally black. Brood cappings are discoloured, sunken and sometimes perforated. There is generally a distinctive odour associated with the presence of disease, but it is necessary to have witnessed it for that to be a useful diagnostic tool. Authors have variously described it as "not strong or unpleasant" and "distinctive odour likened to dog manure".

Since 1907 it has been known that the causative agent is Bacillus larvae, but it took another thirty years to discover that only the spore stage of the Bacillus encouraged the disease. Honey is the vehicle for transmission. In the U. K., American Foul Brood is a notifiable disease and therefore, the outcome is not in the hands of the beekeeper. Normally results in the destruction by fire of all bees and equipment involved. The use of antibiotics is questionable, but at best they only save the bees and they were not available in Cowan's era. Before antibiotics, many 'solutions' involved the separation of the bees from their honey. The most popular method involved creating an artificial swarm on new foundation in a new hive, and then confining the bees until they had consumed the honey that they were carrying; the original hive, in totality, being destroyed. This is considered unreliable, only the bees are salvaged, and there is the risk that it will not contain the outbreak.

European Foul Brood.

European Foul Brood is a disease of unsealed brood. If inspection reveals the occurrence of gaps in what should be solid blocks of brood, it is probably an indication that the colony has European Foul Brood. The smell ranges from diabolical to non-existent and is therefore not reliable as a means of diagnosis.

The cause is the streptococcus pluton, which does not form spores directly, but provide the means for other microbes to invade the infected grub and multiply. The most common of these is Bacillus alvei, long believed to be the cause of European Foul Brood. The larvae swallow the cocci, which then reside in its gut and consume some of the food intended for the grub. If the number of cocci swallowed by the larvae is large in proportion to the amount of food supplied, it will die by starvation and will not be able to pass on the infection. The bees then clear out the remains of the larvae. It is the larvae that are fed adequate food in proportion to the cocci in its gut that survive and have the ability to propagate the disease. The larvae has no means of passing the cocci through its body prior to pupation but when it pupates the gut opens and the contents are spread over the inside of the cell and would normally then be covered by the cocoon. A weakened larvae is not capable of spinning a full cocoon leaving some of the faecal pellets exposed, and it is this that provides the means for spreading the infection by the house bees when they clean up.

The level of infection varies through the year, low at the start of breeding and building with each brood cycle, but it is very dependent upon food supply. Bad weather and cessation of nectar gathering will result in an increase in the number of larvae deaths, and thus reduction of the infection level. If a cure is successfully applied, re-infection is possible when the bees break down a previously infected comb, thus exposing some of the pellets previously sealed away behind the cocoon silk.

European Foul Brood weakens a colony but rarely kills it, and it has the ability to remain undetected for many years. Whilst it is always advisable to maintain stocks in as strong a condition as possible, it is not a guarantee that such colonies will not have E. F. B. Robbing, drifting and beekeeper action is the most common methods of spreading the disease.

Treatment is by means of antibiotics, but in the U. K., E. F. B. is a notifiable disease and hence the outcome is in the control of a local Bee Inspector.

However, in the early decades of the twentieth century, beekeepers' focus in Britain was diverted to the new, much bigger problem that was Acarine (Isle of Wight) disease, but legislation would apply to all diseases.

Background

The early issues of the *B. B. J.* were concerned mainly with instructing its readers on the modern methods of keeping honeybees. C. N. Abbott, the editor, was no authority on bee disease. However in the fifth number he wrote of foul brood in his editorial, saying that it was zymtopic, highly contagious and the honey was charged with the germs of infection. He added that it was not observable in unsealed brood probably because the bees remove it. His solution was a version of the purgatorial method, - place bees of infected hive in empty hive on foundation, confine, feed for a few days until the honey in the honey sacs was converted or consumed, then unite with healthy stock. Boil honey, burn or bury hive, brood and all combs.

It was the start of many claims for cures, and disagreements on how to resolve it, but all lacked the necessary scientific input. J. S. Wood. Nyborg, Denmark, wrote criticising some of the contributions on the topic and followed by translating a paper by Pastor Schonfield (German), in which he claimed to have proved the theory of Dr. Preuss (Prussia) that the spores of the micrococcus travelled in the air from the putrefied brood. He claimed to have proved it to be infectious, the infection being produced and carried by micrococcus. Wood prescribed Salicylic acid as a solution to the problem, describing his method of using it. At this time Salicylic acid was used to preserve fish, meat, etc. by preventing the growth of mould. So what little research or investigation that was taking place, was happening outside Britain by beekeepers with a scientific bent. It was no surprise that it was German beekeeping scientists leading the investigations; Germany was the first to recognise Science as an academic subject in its education system.

Whilst the B. B. K. A. held quarterly 'conversaziones' during which papers were read and discussed on many things, they did not turn their attention to foul brood until early 1884. Samuel Simmins, a commercial bee man, and paper proprietor, was chosen to read a paper on brood disease. It was a strange choice because the only individual in Britain who was carrying out any kind of scientific work on the disease was Frank Cheshire, a retired schoolmaster, but he had fallen out with Cowan and the B. B. K. A. In his paper, Simmins believed that foul brood was brought on by neglecting chilled brood, and had been induced by keeping the combs too long in the extractor when removing the honey in them. His solution: destroy combs, reduce bees to state of starvation, then place them on new comb foundation in a clean hive. He said that it was infectious and spread by robbing or beekeepers. He did not believe that there were two forms of foul brood as Dr John Dzierzon of Silesia had reported. Cowan was not at the meeting but responded at length with a letter to the *B. B. J.* as follows:

```
When he (Simmins) tells us that the only method of curing
the disease is by destroying all the comb and brood, I am
sorry I cannot agree with him…….with our present knowledge
of the disease we have the means of effectually curing it in
a rational manner…..much less cost than would be involved
in the wholesale destruction recommended, pointing out that
this would be ruinous to those with a large number of
hives.   ……I have considerable experience with foul brood
and had it at a time when not so much was known about it
as there is now ……..have tried all and there is no known
remedy superior to the salicylic acid treatment.
```

He gave the formula for syrup containing salicylic acid, and mocked Simmins for being imprecise – "a spoonful" being no scientific measurement at all. This would return to bite him later when a correspondent criticised him for exactly the same lack of detail in a formula in his *Guide Book*. Often Cowan's articles include very strange sentences, – "Although I believe that chilled brood is amongst the causes of foul brood, I entirely disagree with him that it will in time create the disease or that it can even do so". He praised Hilbert's fumigator – he had seen it in France and reported on it then. In the discussion Mr Zehetmeyer denounced Salicylic acid, because "on the basis of two or three hives treated, it did not work for him". In his usual fashion Cowan says that his failure must have been because of the way he did it. Zehetmeyer recommended Thymol. Cowan thought that it was disagreeable to the bees and more dangerous to handle, but did not say that it did not work.

Whilst the B. B. K. A. were not prepared to hear Cheshire, his time came when he presented to a large audience at the International Health Conference in Brighton later the same year. He commenced by relating that it was normal for beekeepers to send comb samples to the offices of the *B. B. J.* asking if they contained foul brood. The editors duly obliged in their 'responses to queries' column. One such correspondent had halved his sample and sent one half to the *B. B. J.* and the other to the *Journal of Horticulture*. Cowan/Carr responded saying that it was dried pollen and the *Journal of Horticulture* declared it foul brood. As would be anticipated, the *B. B. J.* editors ridiculed the report in the other magazine. Cheshire said that he was responsible for the *Journal of Horticulture* diagnosis – the one person in Britain at that time able to accurately recognise foul brood! He proceeded to give details of how he had inspected 200 samples sent to him, described the difference between micrococcus and bacillus, and identified the bacillus under the microscope, but questioned whether it was cause or effect. His studies proved Schonfield was in error by calling them micrococci, and that foul brood was a bacillus disease – the 'germ theory' was pre-eminent at the time. He claimed to have seen it eight years earlier, but there is no published evidence to support this, and it probably owes much to the fact that 'his' discovery was in another author's book of that date.

He continued by saying that he was pleased that Cowan had not followed the crowd when he said "foul brood does not *seem to* affect the bees", however, he then said that he had found active bacilli in the stomachs of bees, and it would be logical that queens also carried the bacilli. This led him to conclude that the importation of bees from Italy was the method of introduction into Britain, something Cowan was to repeat many times. This was a desire to blame someone else for the problem and does not concur with history. Cheshire believed that once established in a hive, the antennae of the nurse bees were the medium of transmission, later adding feet and mouth, but he had found no live bacillus in any honey, so that could not be the method of propagation. In 1878 Cowan wrote "….the honey which is supposed to contain the spores, although I must say that I have never been able to detect any by the microscope." Cheshire named it Bacillus alvei.

Cheshire continued by disagreeing with the formula used by Bertrand, Hilbert and Cowan, and thought that the fumigators of the first two would not be effective. He recommended using Phenol acid in place of Carbolic acid for disinfecting hives, and Salicylic acid should only be used in the feed. He finished by attempting to divert some of the comment he was certain would come, by saying that years ago he and Cowan stood alone in recommending curative measures because they had both cured it.

There were many follow-up letters pointing out that he was not the first to spot that bacillus was the cause, and the use of Phenol was not new, it had first appeared in the *Journal of Horticulture* in 1865. Cheshire had not claimed either, but he was the first to name the bacillus. Watson Cheyne, understudy to Lister of bacteria fame, joined Cheshire in the research work, giving it considerable gravitas. The pair read a paper to the Microscopical Society a short while later but it added nothing of significance. However, credible research was now taking place in Britain to, at least, add to the multiple cures being advocated by amateur beekeepers, all believing that they had cured it.

First attempt.

The B. B. K. A. continued in its, by now, familiar fashion: very formal meetings, organisation of the occasional show and talking endlessly at the quarterlies. In the Spring quarterly of 1885, they talked of foul brood again and unusually, ended by passing a resolution requesting that the B. B. K. A. Committee determine through secretaries of local Associations and experts, the existence and extent of foul brood in their locality, and when they have that information send it to the B. B. K. A. who would then communicate with the Ministry of Agriculture on the subject. The Committee discussed it "at some length" and then did exactly as the resolution suggested, but there were no directions given as to how the local secretaries were to obtain the information, the amount of detail to be included, or the form in which it was to be presented. So there could have been little surprise that the information obtained was not 'meaningful' and not good enough to be put to the government. Judging by the published material available, no disappointment either.

The lack of enthusiasm amongst the Committee members might have been due to Cowan's absence for the complete summer, but for him it was something of a dichotomy. He believed that "all advanced beekeepers" could cure the disease using his methods and the nostrums that he sold. However, he recognised that legislation would involve inspections by 'experts', and the B. B. K. A. had now added a separate part to their expert's certification on foul brood. He saw this as the only source of expert advice available, and being chief examiner, legislation would add to his power and importance. So with legislation, government money would be channelled through the B. B. K. A. to operate the scheme and into the pockets of Cowan's experts. A fee of two guineas per day was the suggested level of remuneration, about £300 today. He responded to those who criticised the double standards by saying that those who followed his doctrine had no cause for concern because they would not have the disease: a win-win situation then!

The inability of the central Association to collect the relevant data from bee-keepers resulted in this first attempt at obtaining legislation ending in failure, if this could be called an attempt. The next logical step would have been for the B. B. K. A. to analyse why they failed to obtain the statistics and work out a strategy to ensure that they had them in the future. After all, they were the bee-keepers who had said that the disease was widespread in the country, requiring legislation to deal with it; but it was their arrogance that prevented them realising that the Government was unlikely to provide money on the basis of them just so saying. The shock should have produced action, but for a group of London gentlemen bee-keepers that was never going to be the case. They were not even capable of devising a template for presentation of the results from local Associations; and how would the average bee-keeper know if he had foul brood?

Second attempt.

The problem was going to go away. Bee-keepers were still being vocal seeking some form of compulsory powers for the control of foul brood. Other Countries were already introducing legislation, often quickly with little opposition. Pringle, a Canadian bee-keeper that Cowan had met in London and Ontario, wrote detailing their scheme. When it was published in the *B. B. J.* the editor added the footnote "would that England had a similar law". This was an association that upon his return from visiting Canada, Cowan declared was "not as good as ours"; but clearly more effective.

The Chairman of the B. B. K. A. placed the blame for failure on the local Associations, their secretaries in particular. Some of them had conveyed the information that on the basis of their local knowledge foul brood was practically unknown in their county. Cowan thought that he knew better, based on his and Broughton Carr's inspection of pieces of comb sent to the *B. B. J.* offices. The *B. B. J.* had received considerable correspondence criticising the B. B. K. A. for the failure. Cowan responded with a series of 'rants' at local Associations. In one such editorial he published a letter from "one of those who criticise without taking their share of the work", contradicting his often repeated claim that it was necessary for the Committee to comprise members from the London area, because others could not take their share of the work. It would have resonated better coming from someone who had not just returned from a very long holiday in Switzerland.

The letter Cowan attacked was from "A. D." and headed "Observations by an outsider".

> It appears to me that the authorities on matters connected with beekeeping in London (i. e. the B. B. K. A.), are by degrees acquiring an exaggerated estimate of their own importance and endeavouring to arrogate to themselves powers stronger than the moral influence they have hitherto made use of with so much success.

He believed that the reason for seeking legislation was to:

> ...take decisive steps for its extirpretation ......exercise of repressive measures will be a blow to British beekeeping.

He believed that it is the cottager that would fear this most, and continues by saying that he has not seen foul brood even though he has handled many hives. He then questioned if this ignorance would disqualify him from the new expert certificates that now contained a section on foul brood. Following this letter it would be a waste of time applying. He continued "... I fail to see any real value in the certificates". As is now anticipated the editorial response was long and detailed, and of course the only reason for publishing it.

In his response the editor states that he knows that the correspondent's county is riddled with foul brood, and most beekeepers there are struggling to carry on because bee-keepers with diseased stocks refuse to destroy them:

> Surely this is 'collateral evidence', sufficiently strong in support of our conclusion. Does any sane man......possess so limited experience of the disease as to be unable to tell when he sees it.

Cowan then implied that "A. D." rated himself an expert, but that could not be the case because he did not make himself acquainted with the nature and characteristics of foul brood and its means of possible cure. He also included more of the usual "Surely it is patent to anyone....". style rhetoric.

So, not for the first time, there was considerable disquiet in the ranks of bee-keepers. Local Association secretaries were not happy with the public beating from Cowan because they had not been able to do the impossible; the B. B. K. A. and its Chairman were not happy at being criticised for lack of, or too much, authority and too little activity and bee keepers in general for not following the teachings from the top. There was lack of scientific research taking place; lack of unity amongst beekeepers as to the way forward; lack of knowledge of the disease by most beekeepers and lack of any statistics to provide support for any approach to the government.

The *B. B. J.* marketed a photograph, taken by Cowan, at 1/-. An engraving of it appeared in the magazine, but little else of consequence was happening. To hold a democratic vote of beekeepers to determine the support for legislation, I do not believe entered Cowan's thoughts at this time, and it was definitely beyond the capability of the central body.

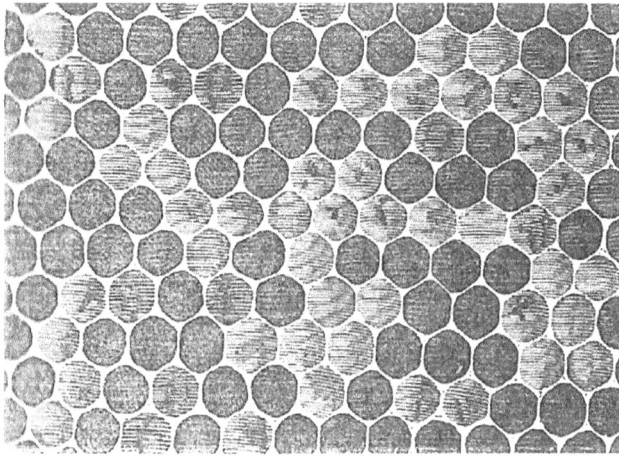

Cowan's photograph illustrating foul brood

By the early 1890's for the first time public money was being channelled through Local Authorities for Technical education, some of which was being allocated to bee-keeping often in conjunction with the County Associations. Many approaches by the central beekeeping body to the Government for funds had failed, being informed that they would only be allocated for specific educational objectives. It took the B. B. K. A. several years to come to terms with this; certainly the Chairman was of the opinion that all money for bee-keeping should be channelled through his Association. This was the clearest indication yet, that any approach by the B. B. K. A. to the Government in respect of legislation would result in a scheme being operated at local level through the County Councils in conjunction with County Associations.

Many editorials, letters and meetings later the B. B. K. A. sought to include foul brood among bees in the Contagious Diseases (Animals) Act, naively claiming that this would impact little on the public purse but, of course, they had no idea of what the actual financial implications were. This appears to have stiffened opposition to compulsory powers as the B. B. K. A. were now calling legislation. The Association and its Chairman were being made to look ineffectual, especially to those countries where powers were already in place. Cowan was becoming very agitated by the continued opposition to his stand, his language becoming ever more irrational in his editorials. "......sinks of contagion, should be destroyed and put out of the way of doing harm whether their owners like it or no". He called for concerted support for the B. B. K. A. in this new approach to the Department of Agriculture. This new approach was to highlight that the public money now going to bee-keeping education would be wasted if Government does nothing to control foul brood. This approach was ill thought out, a cornerstone of any legislation would require bee-keepers to be educated to recognise disease.

As would be expected the opposition was not silenced, their doubts being met with " Surely it should strike the most inexperienced among us that no man in his senses would dream of ....", the sort of rhetoric that was always sure to win over the doubters! In the latter half of 1894, the *Journal* editorials on foul brood and related correspondence continued relentlessly, one anonymous correspondent bizarrely claiming that compulsory powers were a method of introducing socialism; guaranteed to scare the central Committee members.

In December the B. B. K. A. responded and set up yet another committee "to consider the matter of foul brood as an infectious disease and to submit the same for consideration of the Ministry of Agriculture". Although Cowan was not at the meeting that set up the committee he was inevitably a member of it. Early 1895 Cowan called time on correspondence on legislation, stating that it will now be left to the committee appointed.

The committee drew up "suggestions" to be incorporated in a Bill; the Irish B. K. A. requested that any measures be extended to Ireland, which was agreed. Cowan and another member of the committee, E. D. Till, met with the Secretary of the Board of Agriculture, which resulted in the President of the Board of Agriculture, Gardiner, agreeing to receive a deputation on the matter. They were "most cordially received, much valuable advice being given them". This valuable advice was again not shared with the general bee-keeping public. Till had only kept bees for a couple of years.

The passage was not as smooth as Cowan would have us believe from his jubilee history articles. Many wrote directly to the Board of Agriculture, objecting to compulsory powers, or to some of the 'suggestions'; they included a Council member who gave details of the cost of the scheme for Kent.

Hewitt was doing his usual thing in the *Journal of Horticulture* pointing out that the B. B. K. A. experts used bees with foul brood in demonstrations, and were still demonstrating driving using skeps, which were a prime hazard; thus they were deliberately spreading the disease. When a correspondent wrote to the *B. B. J.* quoting the article, the editor's lengthy footnote included:

> The whole of the statement quoted above is so thoroughly improbable that it refutes itself…….. it would be beneath the dignity of the gentleman who leads the B. B. K. A. to treat such allegations with anything but the silent contempt which they deserve.

Easier than addressing the points raised.

The deputation took place in May 1895. The *B. B. J.* carried a full account, comprehensive list of those in attendance, and every word spoken. The B. B. K. A. had eleven representatives, and there were a large number of M. P.'s and others present. After the introduction Cowan was the first to speak, with even the 'hear, hears' being included in the *B. B. J.* report. It was of course usual. Several others spoke in support. Ultimately Gardiner responded at length with an obviously prepared statement raising certain 'difficulties':

- Are there any preventative or remedial measures possible? He wants to hear from scientific beekeepers.
- Assurances were needed that all beekeepers were in favour of the action proposed. He had met trouble with other such cases. Statistics were required to prove the case.
- Would an "honest and careful" beekeeper be able to recognise the disease? Vital, because compulsory notification would be essential for any legislation to work.
- What assistance would be required to check the accuracy of a reported incidence? Important to avoid fraud if compensation was involved. Bee-keepers cannot be trusted! Expert assistance had been quoted at two guineas per day.
- Evidence required that the benefits of the scheme would outweigh the cost of implementing it. Suggests costs could be limited by the B. B. K. A. experts acting voluntarily, or for small fees and expenses: not exactly what the experts had in mind.
- What level of compensation was to be set?
- It would have to be an adoptive Act if it was passed.

Cowan was well pleased with himself. It was a "large and influential" deputation, and some newspapers had long reports of the proceedings, his speech being reported ver batim, whereas others had been summarised. He would repeat the importance of this meeting many times.

However, it is difficult to view it as a success. The B. B. K. A.'s case had been seriously damaged by 'difficulties' most of which could have been foreseen, and responses included in the presentation. The arrogance of those involved, especially the Chairman, is astonishing; he believed that the case he had made would be accepted simply because he had made it! Gardiner did not suggest that Cowan was lying, just that he needed evidence to support the claims being made - "…the unanimity with which beekeepers were welcoming the prospect of measures being taken for dealing with it" not good enough, then. It was a very amateurish attempt, and the eventual outcome was not in doubt because most of the 'difficulties' could not be resolved. Since when have beekeepers been able to universally agree about anything?: three bee-keepers, two problems, equals six solutions. In his account on the history of the B. B. K. A., Cowan 'glossed over' Gardiner's 'difficulties'.

The wheels continued to turn, the committee revised the suggestions for inclusion in a Bill; another meeting with the secretary of the Board of Agriculture where "….all points raised were dealt with and answered, it is confidently believed, to the satisfaction of the Department" but again no details published.

A major point in Gardiner's response to the deputation was that the County Councils would administer any Act, and it would therefore be 'Adoptive'. It would be for each County Council to decide to adopt it, or not, and operate the scheme. The B. B. K. A. reluctantly decided that it was necessary to engage with County Councils and convened a conference to which all County Councils were asked to supply representatives; result, much talk and another committee. A revised draft Bill was produced and taken to the Board of Agriculture where they were informed that there would be no bill that session, and it was suggested that they get a private member to introduce it, logically the Chairman of the County Council Committee – Lord Thring; parliamentary language for 'go away'.

It was now early1896. The *B. B. J.* published the eventually agreed Draft Bill; great fanfare; sent a copy to the Board of Agriculture who gave "…..a favourable reply…..containing valuable suggestions", so still not correct. The *B. B. J.* was now full of the new found friend, Lord Thring, and their dealings with him, but when the time came, he failed to get the requisite signatories needed to introduce the measure, and the second attempt ended in failure.

The sole positive outcome was that the Board of Agriculture agreed to publish a leaflet on foul brood and distribute it free of charge. At least it might go some way to solving the problem of education and recognition.

We are informed that this whole exercise had cost the B. B. K. A. in excess of £100, which was covered by individual contributions, but are supplied with no information on how they managed to spend such a large amount, or who donated. To put it into perspective a copy of the *B. B. J.* cost 1d (240d = £1) and the average weekly wage for an agricultural worker was about 12/- (£0.60). Cowan later claimed that "The failure was greatly due to the reluctance of bee-keepers to supply the necessary statistics". Whilst statistics on the incidence of foul brood would have been unreliable, those for or against compulsory powers would have been better, but bee-keepers had not been asked. Gardiner had raised many problems that the B. B. K. A. seemed unwilling to address, and opposition was much stronger than Cowan was prepared to acknowledge. But the Chairman was now off to America for a 'lengthy sojourn'. Was he trying to avoid having to face criticism over yet another failure?; we shall never know.

I have detailed the next part of Cowan's life and his movements elsewhere, but I summarise below his major travels for the next nine years.

1896 November to America

1898 June to Britain

1898 October to America

1903 May to Britain

1905 April to America

1905 November to Britain At this point he was close to sixty-six and retired to Somerset.

Third Attempt.

In his history of the B. B. K. A. Cowan claimed that:

...the council of the B. B. K. A. were in close touch with the Board of Agriculture (during his absence) ...... but a suitable opportunity (to bring forward legislation), had not presented itself.

There is no published evidence to verify this claim. Upon his return to Britain in 1903, and perhaps the reason for it, Cowan found the B. B. K. A. under pressure, especially from the affiliated Associations. He chaired the October 1903 meeting with Association representatives, arranged as usual to coincide with the Dairy Show. He opened, attempting to immediately divert the meeting, by proposing foul brood as a candidate for discussion saying that:

A few years ago the parent body and its affiliated Associations were in close touch on the subject of foul brood but since that period they, unfortunately, had to some extent lost connection.

He did not succeed. There followed criticism of the B. B. K. A. Basically they did little for the local Associations and had allowed other countries to progress at a quicker rate. A letter was read from George Saunders, Douthwaite, Keswick, Cumberland, Hon. Sec and Treasurer of the relatively new Cumberland County Association. He wrote:

> Dear Sir, — Not being able to attend the Conference to-morrow will you kindly bring this letter before the B.B.K.A. Council. "I would suggest that examiners for expert certificates be appointed at various centres all over England, and thus save considerable expense to counties distant from London, and lessen the feeling which is growing, that the B.B.K.A. would be better named the London B K.A. Surely any bee-keeper who has passed his second or third exam, should be capable of taking these examinations. Such an alteration would tend to make more beekeepers go up for examination, and thus indirectly help bee-keeping generally.
>
> I have had some correspondence with the Irish Bee-keepers' Association, and they are making great efforts to get a Foul Brood Act passed this Session. The Irish Board of Agriculture is moving in the matter, and they seem to have considerable hopes of getting an Act passed. Will you kindly inform me if the B.B.K.A. are joining in this?
>
> I do not think that simply because it has been attempted before and failed there is any reason for not trying again; but just the reverse, I believe more in a policy of ' We won't be happy till we get it.' And that is how the Irish have got so many concessions and advantages to their beekeepers, which we in England have not, from the Government and County Councils. — Yours faithfully,

There was no immediate response to this letter. Others continued to pile on the criticisms: -

- B. B. K. A. should be renamed the London B. K. A.
- Somerset B. K. A. complained about scant courtesy it received to a communication, and their follow-up was totally ignored.
- B. B. K. A. Expert W. Herrod only had second class certification
- Who examined or appointed the original examiners?
- As would be anticipated the Chairman responded robustly: -
- No point in electing anyone from outside London they did not attend and take share of work
- B. B. K. A. very happy with W. Herrod with or without first class certification, they were sure he could "pass the examiner."

- The examiners were initially chosen "as those supposed to know about examinations". He gave four names, but one individual took all of the early examinations. One of the names he gave was not a member of the B. B. K. A. at the time of the first examinations, but he was a schoolmaster and hence meeting the criterion better than the others. Who were there at the time of the first examinations to examine those appointed anyway? He "challenged the impeachment of that system".

There were other criticisms and reposts but no mention of foul brood, the ploy had not succeeded and he was too busy fighting off criticisms of his Association to remember where he started. Cowan finished by relating, tales of American beekeeping requested by one of the faithful, - much less controversial and nobody seemed to care.

Saunders had to wait until late January 1904, before he received a response to his letter. The B. B. K. A. Council instructed the Secretary to reply to the effect that they had good reasons for knowing that such an Act could not possibly be passed at present. But Saunders was not about to give up. Cumberland B. K. A. had employed an expert to tour, inspect and deal with the foul brood that he found. The County Council had allocated funds to the Association for this purpose. Saunders saw that the way forward was for all County Associations to do the same as he had done in Cumberland, then join together, with the facts thus obtained, to approach central Government seeking the necessary powers by way of legislation; basically the same as that proposed by Gardiner. Because of the indifferent response from the B. B. K. A. Saunders next approached his local M. P. who wrote to the now President of the Board of Agriculture, Lord Onslow, who responded saying that he would definitely pass a foul brood Bill as a Government measure if County Bee-keeping Associations could get their County Councils to ask him for such a Bill.

The 'Saunders scheme' stood a better chance of success than anything that had preceded it, and was a more professional way of operating than a few English gentlemen claiming to represent all bee-keepers, approaching the Board of Agriculture and presenting them with a draft Bill. The problem was, of course, that the B. B. K. A. had no role in the Saunders scheme. Although he had effectively been rebuffed, Saunders continued to keep the B. B. K. A. informed, urging them to co-operate. Several County Bee-keeping Associations agreed to support the measure, as did the Irish B. K. A.

In an attempt to discredit Saunders the April 1904 edition of the *Beekeepers Record* carried an editorial on foul brood, concentrating on his efforts. There was much praise for the 1895/6 attempt, the efforts of the B. B. K. A. and it's

Chairman. The editor claimed that this new scheme had not reached the level of the previous one and would be very expensive for the County Councils. The article relied heavily on material that had only previously appeared in the *I. B. J.*, but it was unlikely to convince many doubters.

As would be expected Saunders responded with a letter to the *B. B. J.* which Cowan included in an editorial. Saunders outlined the history of his campaign and claimed that the Bill was being prepared "at a trifling cost" and that Ireland wished to join with England. He repeated that he had requested the B. B. K. A. to join but they had declined, so he intended to "endeavour to get the County Associations to combine and pass an act independently of the B. B. K. A." Cowan responded to this selectively quoting letters on the matter from the Department of Agriculture to justify non-action. He yet again, repeats, with much self-praise, the previous efforts, and ends with a challenge to Cumberland and other supporting County Associations to guarantee a small sum, "say £20" to provide the necessary money. It is a childish thing to write, devoid of any logic. Who or what would this guarantee be for? Where has £20 come from? Cumberland had already done it, had it cost them £20?

The next issue of the *B. B. J.* revealed that it:

> ...... had been favoured with a further communication from Mr Geo. Saunders......complaining of the way in which his letter on the subject ...was dealt with ......but the limitation of space precludes its publication just now.

There followed a full page reviewing an obscure Swiss leaflet of 47 pages. Also space was found to include yet another anonymous letter claiming that Saunders action was not representative of those in that Association. Saunders immediately wrote again not accepting the 'excuse' offered. Not published, this time because it contained 'personalities' and needless abuse. However, the editor accepted that a factual error had been included the previous week "under pressure to meet the printing deadline". The same issue contained a report on the B. B. K. A. Council meeting in April, and included a letter to Saunders repeating their position. It is this letter that Cowan included in his history of the B. B. K. A.

The *I. B. J.* for May1904 contained Saunders letters: –

- Lowther, M. P. to Saunders giving Onslows response.
- Saunders letter to Rev. Digges editor of *I. B. J.* giving cost of drafting Bill - £2 4s 6d or £3 5s 6d "should it prove troublesome", the only other costs being printing and postage.
- The first unpublished letter, Saunders to *B. B. J.* including the parts omitted, complains about factual error and that the responses from the

B. B. K. A. secretary are curt; refers to the cost; clarifies point referring to a mention he made about Ireland.

- The second unpublished letter containing the 'personalities and needless abuse' is reproduced below without comment.

To the editors of the B. B. Journal.

Dear sirs, - In consequence of your refusal (as I cannot really take your excuse of "crowded out" seriously) to correct your editorial of April 14th, I must decline to write further on the above subject in your Journal, but I would refer "A Cumberland Bee-keeper" and others interested, to the IRISH BEE JOURNAL, which will continue the subject. It surely shows a weak case, that you should need to quote one anonymous writer, whose identity I believe I happen to know, and who has formed false impressions of facts, solely from your Editorial, and I would reply to him, asking him to kindly reserve his judgement till after the annual meetings, when above will be on the agenda; and those have been delayed by my absence from home through illness. What he calls my "private" views are formed by intercourse with members in the county, and he will find that the present "F. B." Bill measure is not costly. Yours faithfully, George M Saunders.

There are further letters Saunders to B. B. K. A., the reply, and Saunders to *I. B. J.* reporting the correspondence.

For any organisation this is the stuff of nightmares, a loose cannon in the ranks and the great leader making no impact, although he repeated at every opportunity that legislation was not achievable at that time. The danger was, that Saunders would succeed; he was certainly not going away. Relying heavily upon the 1895 effort and legal assistance his Bill was drawn up and distributed to all County Associations, who met to consider how to react. Some, notably Devon led by H. J. O. Walker a long term Cowan ally, immediately contacted their County Council. At the June B. B. K. A. Council meeting, apparently following many approaches on the subject:

......it was eventually resolved ......to ask all County Associations to apply to their respective County Councils asking whether an Act would be supported if it were obtained.

Digges, in his magazine welcomed the move, but added that they should have done it months ago. The central Council hoped that the County Councils would baulk at the cost.

The *B. B. J.* correspondence columns now began to take on a new look. Letters expressing opinions against compulsory powers appeared, giving a more balanced view than previously. A couple of letters from Saunders were published and although conciliatory they still warranted the usual style of footnote. The *I. B. J.* was now limiting its entries on the subject to calling for support in Ireland for the proposed Bill. By summer, peace had completely broken out – Cowan was in Switzerland and his magazines were very quiet.

The topic for the B. B. K. A. Autumn conversazione in October 1904 was to be foul brood. Cowan's editorial in the September *B. B. J.* set out what he thought should be the main points to be discussed. Acting in his position as senior editor of his magazine, he was effectively instructing the organisation of which he was Chairman, how the meeting should go. He points out that there is little difference between the Saunders Bill and his own of 1896. There are a few alterations:

> ….. but some at least can hardly be called improvements. For example the one that extends the scope of the measure to Ireland.

I do not believe the Council discussed this point. The I. B. K. A., the *I. B. J.*, the Irish in general and Digges, in particular, had been an irritation to Cowan for many years, but this was an extraordinary juncture for it to re-surface. He had contrived to make the point. He also strongly states his opposition to compensation, but that also had little logic and illustrated his lack of connection with the 'cottager'.

The Chairman had decided upon the agenda and made clear his views prior to the meeting. At the Council meeting a week earlier it was decided that the large amount of correspondence received on foul brood would be summarised and presented to the Conference on the 6th October. The next issue of the *B. B. J.* contained the Saunders and B. B. K. A. draft Bills side by side to prove the point that there was little difference between them, but the wording of the Bill had not been the reason for previous failures.

The Chairman opened the Conference minutely detailing the previous attempt and stating that "Since then nothing had been done until the recent correspondence initiated by Mr. Geo. Saunders" - not exactly what he reported later in his history. Saunders interrupted to say that he had all the correspondence with him - not what the Chairman was used to when speaking. When Cowan continued he gave details of how every Association had responded to the B. B. K. A. Questionnaire. He was interrupted many times by County representatives correcting what he was saying for their Association. The meeting quickly progressed to the state of shambles. When he eventually finished reading this detail he called for a vote by the delegates, for or against legislation. Had he not just read it out? Undeterred delegate after delegate gave details of their Associations views; they had come prepared to have their say, and were not going to allow it to be simply, and inaccurately, summarised by the Chairman. Saunders said his piece, but when published in the *B. B. J.* it too was 'summarised'. More speeches from the floor, resolution proposed, more speeches, amendments to the proposed resolution, even more speeches, but when eventually there was agreement on the resolution to be put, the Chairman declared that it "would be deferred to the end of the proceedings". Presumably, he had guessed the outcome of the vote and was not happy with it, so best not have it!

Cowan then proceeded to attack Saunders and his Bill, with the same arguments that he had written in the *B. B. J.*, adding that he had been to the Board of Agriculture and met with a junior official who confirmed that Lord Onslow had not given the pledge as Saunders claimed. But Lord Onslow's letter had been published in the *I. B. J.*, it was clearly a conditional pledge, and Saunders was not claiming otherwise, and protested strongly. It is a childish argument over words. Saunders also firmly stated that he could not consent to the omission of Ireland.

The Chairman then 'summed up' by repeating all of what he had said before. It was so negative that it could be interpreted as not being in favour of legislation, but was more to do with his inability to cope with the situation in which he had put himself. Having declared that the whole thing would have to be dealt with by a committee, and failing to put the agreed resolution to a vote, he then 'set about' Ireland:

> ...it appeared to him quite illusive to suppose that the Irish Association could effect anything at the present time. The Irish Bee-keepers Association once a powerful one with a large membership, had, he understood, dwindled down to about one hundred members, so it was not surprising that it has little influence in the present effort for legislation. In fact Mr Wyndham has already stated in the House of Commons that the Department of Agriculture did not intend to ask for legislation until the measures it was taking – of which he (the Chairman), had already spoken – and which were now effectively working, had failed. In point of numbers, the whole of the Irish Association mentioned in Mr Saunders' circular probably did not reach a higher total than some of our larger associations in England; and it was not to be wondered at if their influence was lessoned on that account. He, therefore, felt that, however unfortunate it might

be for their friends in Ireland, our cause would be not
advanced, but retarded, if joint action was adopted. The
best course for English bee-keepers was to ask for their
own Bill, and Irish bee-keepers could ask the Government to
extend it to Ireland.

As usual it contains many Cowan interpretations of 'fact', and would normally have met with unanimous support, but unusually none was voiced; Saunders interjected again correcting, as he saw it, the errors in the statement, ending with "......he was afraid he must continue to work alone, as he was pledged to Ireland."

The Chairman had not finished, and continued with "...he was given to distinctly understand......" and then repeated all that he had said before. He added that again in relation to Ireland, Wyndham (Government Minister), had said that the funds for Ireland "had been hypothecated up to the hilt", and in an attempt to prove the point gave the statistic that the Irish Board of Agriculture had given 719 lectures on bees, poultry and other subjects at a cost of £18,000 – about £25 per lecture, many of which were poorly attended. It was an extraordinary rant, even for Cowan. Many of his statements would be challenged in the future, but he was very much in denial. A motion to establish a committee was put and passed, and unusually three names, one of them Saunders, were proposed for that committee. This was a very unusual step and obviously the Chairman was not pleased, claiming that it would be too unwieldy - it would be when the B. B. K. A. had added four of its own!

The decision to appoint a committee did not stop the discussion continuing. Next was compensation; the Chairman was very much opposed to any compensation, and expressed the firm view that the proposed 10/- per hive destroyed was too high because a diseased hive had no value, the reason he believed that it should be zero. When attempting to assess the cost of compensation, he again managed to bring Ireland into it saying that it was possible to obtain statistics in a sparsely populated Country like Ireland where they were "... collected by the constabulary, but it could not be satisfactorily done here". Several years later his adopted County of Somerset did just that and he was full of praise for it.

Saunders and others wrote to complain of, and ask for corrections to be published, to some of their contributions as reported in the *B. B. J.,* drawing the response that:

......we need only say that our reporter took full notes of
what was said at the meeting and – without being a ver batim
report, which is, of course, out of the question, will

vouch for the general accuracy of his report. We cannot
therefore, burden our already overcrowded pages with.......

But it had been, of course, written in such a manner that it was intended
to read as a ver batim report. Some delegates had also written asking him to
revise his summary of their Associations position, often providing evidence in
support; he responded with a full column reiterating the apparently incorrect
information. It was not conducive to getting the support from the local
Associations that the B. B. K. A. needed, or the statistical information required
in order to progress the case. If the Chairman was not prepared to accurately
report their responses, why should they collect and submit them? Anyone from
the Board of Agriculture reading this would soon realise that the B. B. K. A. was
an organisation in complete disarray.

The 'Irish situation' has been a problem for all British governments for two
centuries. In the early part of the nineteenth century, writer and clergyman
Sydney Smith wrote:

> The moment the very name of Ireland is mentioned the English
> seem to bid adieu to common feeling, common prudence and
> common sense and to act with the barbarity of tyrants and
> the fatuity of idiots.

In 1886, Gladstone had tried, but amidst great intrigue, failed to get a Bill
through Parliament granting Home Rule for Ireland. The latest attempt (1893)
passed through the Commons but was thwarted by the Lords, many of whom
had much to lose. Lord Salisbury was one of the staunchest opponents, taking
the line that if you could not rule Ireland how could you anyone else, but
the age of Imperialism was drawing to an end. With this as a background, in
Cowan's case there is also the very personal element.

The Cowan/Digges dispute came to a head on 20th October and is covered in
a separate chapter, although it is inexplicitly linked with this part of the attempt
to obtain compulsory powers to control bee diseases.

Several weeks after the conference, (November 16th), Cowan wrote a long
letter to Saunders, but it is not clear in what capacity he was acting, part of
which read:

> Your experience of public meetings is evidently very
> limited, as you have utterly failed to gauge the temper
> of the meeting, but unmistakably it strongly resented your
> frequent interruptions and this was manifested during the
> recess when nearly everyone who spoke to me alluded to these
> interruptions and wondered at my patience with you; another
> chairman would have peremptorily called you to order.

No self respecting committee would care to be identified or
responsible for the circulars you have sent out, and for
the misleading statements that have been so persistently
made.

You have already caused sufficient mischief with your
circulars and letters.

Obviously no one had explained to the County gentleman that when the Chairman was speaking the only interruptions allowed were, "hear-hear; applause; loud applause; cheers; and loud cheers", even if you did believe that he was talking rubbish. The editing of the report cannot disguise that the meeting was badly chaired. It was the first time that Cowan had come under such pressure in that role; most times these events were meetings of like minds or none at all. Considering the previously given account of the reporting, can any of it be relied upon? The Chairman's contributions will, presumably, be accurate, particularly the "loud cheers", because in his other capacity he controlled the written output. There were no reported expressions of support for his remarks in respect of Ireland.

It is no surprise that Digges responded in the *I. B. J.* to the attack, as he saw it, on Irish bee-keeping and bee-keepers. The initial instalment appeared in the December issue, and briefly contained the following points: -

-   The insulting remarks made in the editorial prior to the Conference and during the event itself, were calculated to produce ill feeling in Ireland.
-   There was a "strange similarity" between much of what Cowan initially wrote and an article in the *Irish Times* by someone known to be hostile to the I. B. K. A. containing inaccurate statements that had already been challenged. This was probably the source of much of Cowans information.
-   The level of membership of the I. B. K. A. quoted by Cowan are challenged and compared with those of the B. B. K. A. as a proportion of each countries population. The case is strong without the exaggeration.
-   He reports on a meeting between the Irish Board of Agriculture and the I. B. K. A. clearly showing that they would support legislation for Ireland.
-   Fuller inspection of the report on the lecture costs would have revealed that only a small proportion of the £18,000 was spent on lectures and none of these were on bee-keeping
-   Because it was clear that Lord Onslow's offer was conditional upon minimal opposition to any Bill, why provoke the Irish with their large and volatile representation in Parliament

The same publication carried further Saunders correspondence surrounding the Onslow 'promise'. The only firm information on this matter is the letter to Saunders from the Board of Agriculture containing the *conditional* promise, and I am unable to find any information to substantiate Cowan's claim that Saunders or the I. B. K. A. had stated otherwise. Cowan uses his interview with a very junior official to substantiate his view on a point that is non-existent. There was no official documentation of this meeting; the only record is that of Cowan.

This had not been a pleasant spectacle but it was to get worse and one can sense the 'red mist' descending.

The "Draft of a Bill for the Better Prevention of Bee Pest" was approved by the B. B. K. A. Council and published in the *B. B. J.* accompanied by corrections to some of the information given by the Chairman at the Conference without acknowledging that it was a correction, or that it is, in fact, from the Chairman! Saunders continued to refuse to supply the B. B. K. A. with the information he has regarding progress with the County Councils, which they now claim is theirs by right. Cowan continued to write to Saunders a further sample read:

> You have a wonderfully exalted opinion of your importance if you, for a moment suppose that I called to the offices of the Board of Agriculture for the purpose of discrediting you …… Certainly in all my subsequent visits I have had more important business to talk about than of you.

> Your No. 7 circular is an ingenious attempt to put a false construction on what is said, but being only *fulmen brutum* (empty threat), is best treated with the contempt it deserves.

Publicly Cowan had written that Saunders resignation from the sub committee was "much to be regretted".

The editorial in the next issue of the *B. B. J.* again gave details of the differences between the B. B. K. A. Bill and that of Saunders, but it was a thinly veiled means of introducing more to justify the Chairman's attitude towards Ireland. This time he tries to convince the Irish bee-keepers that because Ireland is not specifically mentioned in the B. B. K. A. Bill, it does not mean that it is excluded, but it would be best if we "mind our own business". Could he possibly believe this? It continued:

> Our (the B. B. J. editors), attitude as also that of the Chairman of the B. B. K. A., has not only been entirely misunderstood, but has brought forth a vigorous and hostile article……

He believed that this wrong impression of his position is because of

Digges "unreasonableness", and his use of "erroneous information" and "misinformation", and that it is for the Board of Agriculture to decide if the measure is extended to Ireland and Scotland. It is difficult to understand which parts of "extending the scope of the Bill to Ireland can hardly be called an improvement" or "....illusive to suppose that the Irish Association could effect anything at the present time" or "by including Ireland in the Bill our case would not be advanced but retarded", Cowan believed Digges had misunderstood? Without publishing the actual letter he said that he has written to Digges. The letter actually read:

> Dear Sir, - A copy of the Irish Bee Journal for December has been sent me, and I take your leader with the sensational headings as a challenge to me. I therefore accept it as such, and you must bear the consequences.
>
> The attack on myself I should have considered beneath my notice, but as you have assailed the B. B. K. A. and unrepresented its position, I feel bound to defend it.
>
> Yours Truly
> Thos. Wm. Cowan
>
> Let me remind you of the proverb that those who live in glasshouses should not throw stones.

A strangely constructed letter, once again illustrating that he is completely unable to separate his own personal position from that of the central Association. If it was an attack on the *B. B. K. A.*, then the response should have come from them, but this was a private letter from one individual to another.

Why Cowan felt he needed to continue justifying the statements that he had made, especially with such fatuous arguments that had only just come to mind, is difficult to comprehend. It is possible but unlikely, that some of the 'inner circle' had spoken to him about it; certainly there was no published support. However, it is more likely that the *B. B. J.* had received many letters objecting to his stance, particularly from the Irish subscribers, and this was an attempt to show that they had got it all wrong. Perhaps he should have thought of the proverb that contains 'hole' and 'stop digging'!

At the very end of 1904 the I. B. K. A. passed the resolution:

> That the Editorial statement in the British Bee Journal of December 22nd making an unprovoked attacked upon the Irish Beekeepers Association is erroneous and misleading, misrepresents the facts of the case, and constitutes an unwarranted attack and reflection upon the Committee and

```
members of the Association.
```

Cowan's repost included:

```
......but as the Committee (of the I. B. K. A.), - who by their
recent childish proceedings have made almost a laughing
stock of themselves......
```

There were four more lengthy editorials in the *B. B. J.* attacking Digges, Saunders and the I. B. K. A. which included Cowan's version of the history of the Irish Association. There is no doubt that the I. B. K. A. had evolved through, and survived troubled times, during which there had been manipulation of and by the personnel involved, resulting in Digges emerging in Ireland in exactly the same situation as Cowan in England. What was that proverb about glasshouses? I refer readers to *Beekeeping in Ireland* by James K. Watson for a more informative, and I suspect accurate, account of Irelands Beekeeping history than was being given by Cowan, but he was viewing it through 'red mist'.

Most of these editorials were repetition with minute details by way of support, but as usual they contained many misquotes, inaccuracies and insults. By way of illustration:

```
...... Mr Saunders imagines that any course of action he
may take ...... can result, other than disastrously to the
cause...... That any level-headed man, possessing ordinary
business tact, will fail to see this after perusing the
recent circulars which Mr Saunders is so industriously
sending out......
```

Digges responded vigorously but with dignity. Cowan eventually conceded that the cost of lecture figures that he had quoted at the conference, had been sourced from an un-named writer in the Daily Express, that he had invented the authority for the quote therein, and that he had misreported on responses from local Associations. His claim that *The Honey Bee* had been removed by the I. B. K. A. from the books recommended for reading by students taking beekeeping examinations in Ireland, was also acknowledged as incorrect. The last point here referred to Cowan's claim that in a "crowning act of childishness" the I. B. K. A. had replaced his *Guide Book* and physiological volume by Digges's *Irish Bee Guide,* only correct in the case of the *Guide Book.* Did I mention 'red mist'?

In some respects it is impossible not to have a degree of sympathy with Cowan in this whole debacle. He returned from a long spell in America to find 'his' Association asleep to the danger posed by an 'upstart' from out in the 'sticks' who could possibly succeed where he and they had previously failed.

Logically, Saunders, the I. B. K. A. and the B. B. K. A. should have united and worked together to achieve the result they all wanted, but Cowan and the B. B. K. A. would never be involved in any arrangement of which they did not have complete control, and take credit for. The others wanted to be equal partners. However, I do not believe that it was possible to have predicted Cowan's actions, even if the conspiracy theory is subscribed to.

It is my belief that the real reason for much of the anti-Irish outburst was the emergence of the *Irish Bee Journal* and the publication of *The Irish Bee Guide*. Cowan had tried to destroy the latter in a legal case, but had only been partially successful. In the dispute re legislation he never once mentioned the case, leading to the belief that, other than publication of the 'apology' in both parties magazines, there was also a 'gagging' clause. Whilst the Digges book was obvious competition for his *Guide Book*, especially in Ireland, Cowan saw the *I. B. J.* as a much larger problem. This would impact upon the circulation of his two papers. Two English County Associations had quickly adopted the Irish paper as their magazine, and it had a large circulation in Scotland. Also the I. B. K. A. had phased out their agreement with the *Beekeepers Record,* which had supplied the magazine at a discounted rate to members who requested it. There is some evidence that the print run for the *I. B. J.* was between four and five thousand, but it was monthly, the *B. B. J.* at this time being weekly and the *Record* monthly. However, by far the biggest problem for Cowan, was that the *I. B. J.* provided an outlet for alternative views and material that he refused to publish.

The legislation saga continued to rumble on, no chance being missed by Cowan to attack Saunders, the Cumberland B. K. A., Digges and the Irish B. K. A. Eventually a vote was organised, not by the B. B. K. A., but by two individual subscribers to the *B. B. J.* 'Yes' votes to an individual declared in favour of compulsory powers, and 'no' to another individual against. By mid April 1905 the result was published: put simply, 421 voters in favour, with a total of 4.477 colonies, four not declaring the number of stocks; 299 against with 7352 colonies, 46 not declaring the number of hives involved. In the same issue, Cowan published templates with contents, for local Associations, on how to deal with correspondence from Saunders. Saunders was obviously still active and Cowan was becoming ever more frustrated, but he seemed not to understand this was an insult to the very individuals he was looking to for support. The letter carried a date of 10th April 1905, one day before he, yet again, sailed for America. I cannot believe that he had not seen the results of the vote, so why was he still continuing with it? What does it say about his understanding of a problem when, with total control of the bee media and the parent Association, he managed less than 60% support for his view, and what does the total number of 720 voters say about the circulation of the *B. B. J.*? It was of course, only a small sample of the bee-keepers believed to be active in England at this time.

With the leader again away in America the B. B. K. A. Council continued their deliberations. Eventually, on 12th July 1905, they wrote to the Board of Agriculture and Fisheries enclosing the result of the poll, asking if it showed sufficient support for the Board of Agriculture to proceed with legislation. *Oh yes they did!* The reply was received three days later; - failure number three.

Whatever theory one subscribes to on this latest attempt – conspiracy, fear, gross incompetence, or arrogance, it was undoubtedly a mess, and did little for the reputation of any of those involved.

In his *British Bee-keepers' Association Jubilee* publication Cowan summed up thus:

```
Those bee-keepers with level heads and clear business
perceptions could not only credit the Council of the
parent association with having done all that was possible
under the circumstances, but could realise the fact that,
until such time as bee-keepers showed more agreement among
themselves, and until there was more willingness to share
the labour and necessary expense entailed in carrying out
a propaganda against foul brood, it was mere idle talk to
go on suggesting schemes in the British Bee Journal that
were impracticable.
```

In regard to those with "clear business perceptions" I refer you to Cowan's record in business.

There was now a threat other than foul brood to honey bees. Acarine (Isle of Wight) disease was beginning to take its toll on colonies, and the Government was persuaded to investigate the disease. In 1906 the Board of Agriculture appointed A. D. Imms of Christ Church, Cambridge to investigate the Isle of Wight disease but he made little progress and was replaced by Dr. Malden of the Plant Pathology laboratory, Cambridge. He reported in 1909, attributing the cause to a bacillus attacking the digestive system, and named it Bacillus Pestiformis Apis. Later the Board of Agriculture issued a leaflet on Acarine and its control – basically destruction. As in many other areas the involvement of the State was increasing in bee-keeping.

In 1908 a Bee Pest Act was passed for Ireland, but there was considerable doubt about its effectiveness, and by the second year court cases were taking place – reported in the *B. B. J.,* with no attempt made to disguise their delight.

Fourth Attempt.

By 1910 the devastation that the 'Isle of Wight' disease was causing, and the Irish example, led to another call for legislation in England, ironically started again by Cumberland B. K. A. who under a new Secretary, had balloted 700 members and received only one vote against. This time F. Sitwell of Northumberland accompanied Cowan to the Board of Agriculture where they met Lord Carrington who said he needed evidence that the majority of beekeepers were in favour. This was an easing of the previous criteria. Cowan apparently offered the 700 voting papers from Cumberland, as evidence of his much-expressed opinion, that there was unanimity amongst beekeepers for legislation. Yet another committee was formed. This time Cowan was not a member, but he was to be kept informed.

The 1904 Bill was published for consultation, and modified several times in an attempt to quell the arguments against, but any progress was very slow and again the B. B. K. A. were taking on a task that was not theirs; they were just asked to provide hard evidence of support. Eventually, they asked local Associations to poll members and report back. Meanwhile another organisation's magazine *The Smallholder* published their version called "Apiaries Regulation Act". As would be anticipated the response in the *B. B. J.* was abusive. Rev. Tickner Edwardes soon acknowledged that he was "largely responsible" for the alternative because he found "it impossible to agree that these provisions (in the B. B. K. A. scheme), are sufficient to meet the necessities of the case". The "vital points" on which he differed were (i) the fixed comb hive is being allowed to remain in the Country and (ii) nothing is included to control the importation of infected queens, honey and foundation. There was, as would be expected, a letter published criticising the Edwardes Bill.

It was March 1912 before the B. B. K. A. managed to publish the final draft Bill, and in the same month Cowan and W. Herrod attended a conference called by the Board of Agriculture "with object of finding a common basis of agreement". This was unanimously agreed and submitted to the President of the Board of Agriculture, but as usual, we are not informed of the common basis of agreement. The Bill was given its first reading in July; there was much criticism of it in the columns of the *B. B. J.* that also carried another anti I. B. K. A. editorial. The *B. B. J.* stated that the second reading took place in October but was withdrawn due to pressure of work, however the President of the Board of Agriculture promised that it would be re-introduced in the next session. This was challenged by some saying the Bill never actually had a second reading although it was timetabled. However, P. M. Asquith later said that due to congestion of business, unless it could pass as a non-contentious measure, it would not become an Act that session. Failure number four.

Bee Disease.

A

B I L L

To provide for the prevention of the
introduction and spread of Pests and
Diseases affecting Bees.

Presented by Mr. Runciman,

supported by

Mr. M·Kinnon Wood.

Ordered, by The House of Commons, *to be Printed,*
1 *April* 1913.

LONDON:
PUBLISHED BY HIS MAJESTY'S STATIONERY OFFICE.
To be purchased, either directly or through any Bookseller, from
WYMAN and SONS, LTD., Fetter Lane, E.C., and
32, Abingdon Street, S.W.; or
H.M. STATIONERY OFFICE (Scottish Branch),
23, Forth Street, Edinburgh; or
E. PONSONBY, LTD., 116, Grafton Street, Dublin.

Printed by EYRE and SPOTTISWOODE, LTD.,
EAST HARDING STREET, E.C.,
PRINTERS TO THE KING'S MOST EXCELLENT MAJESTY.

[*Price ½d.*]

[Bill 78]

Front page of the 1913 Bill

On the disease research front a team from Cambridge headed by Graham-Smith issued a lengthy report stating that the cause of the 'Isle of Wight' disease was Nosema apis, and named the disease microsporidiosis, but there was considerable doubt that nosema was the cause of bee deaths attributed to the 'Isle of Wight' disease and further research was suggested. Dr. Rennie of Aberdeen University was appointed to carry it out. The Development commission and A. H. E. Wood, a director of B. S. A. and Daimler, who had retired to Glassel in Scotland, jointly funded the research.

The B. B. K. A. tried in 1916 (mid WWI), to get the Board of Agriculture and Fisheries to take an interest in restocking in the wake of the 'Isle of Wight' disease but it was not until 1918 when, following a survey showing only approximately 32,000 stocks of bees remaining in England and Wales[8] that action

---

8   Compare with fewer than 12,000 stocks included in the *B. B. J.* poll earlier, the number quoted by Cowan when addressing Canadian and American bee-keepers and those given by the Government for pre WWI of 78,000 stocks.

was taken, and William Herrod-Hempsall (ex Herrod) appointed as bee expert at the Ministry of Agriculture and Fisheries to carry out re-stocking. By 1920 the Government considered restocking had been largely achieved, although they had only sold about 2,700 queens at 3/9 (19p), each, and 1,600 colonies at £2. W. Herrod-Hempsall was then appointed Technical Advisor to M. A. F. F. and they set up a government service for the examination of diseased bees, operated by Dr. Helen Goodrich of the Department of Comparative Anatomy at Oxford, an arrangement that did not last long.

Dr Rennie announced that he had discovered the Acarine mite, which he initially called Tarsonemus Woodi.

Fifth Attempt.
1920 saw the start of another attempt to obtain legislation. Cowan, now eighty was not directly involved, but the B. B. K. A. Council said they would consult with him on any proposed developments. By now the B. B. K. A. itself was just an onlooker.

The Ministry of Agriculture convened a meeting on 6th Feb 1920 to discuss legislation. Interestingly, they had used the sugar sales for bees made during the war to obtain some idea of the strength of bee-keeping. The figures quoted were 18,000 beekeepers, 64,000 stocks in moveable frame hives, 14,000 in skeps and others. This was estimated to be just 10% of those pre 'Isle of Wight' disease. The attitude of the Government was outlined, and the proposed scheme, which had been handed to delegates as they entered, described. Cowan, who was the first to contribute, complained about not having a copy of the scheme earlier, then:

> I am in favour of the scheme, as it is proposed that the Ministry should undertake the whole management of this thing. ......
>
> Now with regard to compensation; that is a question, I think, to be discussed. I myself am not in favour of compensation. I think bees that are diseased are not worth anything, and the sooner they are got rid of the better. It is for the benefit of bee-keepers and for the neighbourhood, I do not think any compensation should be given. But if you are going to destroy hives and appliances that can be disinfected, of course, in that case, I should certainly say that you ought to have compensation, but not otherwise. I think, with regard to the way of working that it would be much better, as has been suggested, to work it through the

Ministry, and with the Agricultural Education Committees of the different counties. We are doing that in Somerset. The Agricultural Education Committee is in thorough sympathy with the Somerset Bee-keepers' Association, and we have formed, from the Horticultural Sub-Committee, an Advisory Bee Committee. The County Agricultural Education Committee have funds that Bee-keepers' Associations have not, so in that way they are able to help us considerably. As I say, we have formed what we call an Advisory Sub-Committee, which consists of certain members of the Somerset Bee-keepers' Association and certain members of the Horticultural Committee. That, so far, has been working very well indeed, and through that channel we have been able to re-stock a great many apiaries in the county.

Why we want legislation is this. I can give you an instance in our own county of what has been done quite recently, this last year. We supplied nuclei with queens to several of those who had applied for them, and we issued leaflets, larger leaflets than these, with very full and detailed instructions on how to manage bees and how to disinfect hives, and what they were to do. Notwithstanding that, one of our experts went round and found one of the nuclei was being put into a hive that had not been disinfected, after giving them these pamphlets. Now, what are you to do with a person like that?

The Chairman : Disinfect him.

Mr. Cowan : The Chairman says, disinfect him. We should like to do so, but we have no compulsory power. We cannot go and tell him, " You must disinfect", so there we are in a difficulty, and in that way, of course, it will be some time before we can really say we have got rid of the disease in the county.

I do not think I need say much more about it, except that I quite approve of the step that has been taken. I think all these clauses, after they have been thoroughly considered, will be approved, and I think we are very fortunate in getting the Ministry to undertake the matter at last. I have been agitating this for years, so has the British Bee-keepers' Association, and for years we have been pointing

out the danger of these ignorant bee-keepers, and asking
for powers to do something towards stopping the spread of
disease, and I congratulate the Ministry on having taken
the matter up seriously. I am very glad to see such a large
meeting of bee-keepers and other representatives here today.

It would be his final public say on the matter, and was much different than his approach during earlier attempts. Mellowed, or just could not care any longer?

This time the Bill was introduced into the House of Lords by Lord Lee of Fareham, after which it was introduced in the House of Commons. Amid a clamour to reduce Government spending it was again dropped. Failure number five, if indeed it can be considered an attempt.

Cowan did not witness the introduction of Bee Legislation for England and Wales during his lifetime, but he had been involved, at varying degrees, for over forty years and five failed attempts. 1937 saw the introduction of the Bee Disease Insurance scheme, but it was only a substitute for the real thing which occurred in 1941 as 'Prevention of Bee Disease Order 62BB under the 'Statutory Rules and Orders 1941; Emergency Powers (Defence) General Regulations', and was followed in March 1942 by the 'Foul Brood Disease Bee Order'.

Summary

This represents yet another bizarre series of events in the life of my subject. As usual it is impossible to understand some of his thought processes. There is little doubt in my mind that Cowan initially thought that once the Bill he had written was passed into law, as simply a matter of course, compulsory powers would be operated by his Association, including the funding for it provided by the Government. Power and financial reward awaited his friends. This was the initial catalyst for opposition to the scheme, and possibly provided a measure of his own growing unpopularity; not that he would have appreciated that. As each successive attempt failed, the involvement of the Government, initially through the County Councils, increased as that of the B. B. K. A. diminished, and by the time legislation was established the B. B. K. A. was totally sidelined. Cowan's interest reduced with each attempt. How the B. B. K. A. would have managed if it had been obtained at the first attempt is difficult to comprehend. When this started in the 1880's they were struggling to run a couple of shows each year.

I gave considerable space to the 'Saunders' episode because it illustrated, probably for the first time, an insight into how well Cowan could cope with pressure. The answer is badly. Saunder's approach was that preferred by the President of the Board of Agriculture. The County B. K. A. with evidence of problem, and support for legislation, would approach their County Council who in turn would put it to their parliamentary committee, and when there was adequate support they would approach the Government for the legislation. Funding and powers would travel the same route but in the opposite direction. Cowan's preference was, as the representative of a small number of gentlemen beekeepers from the London area, to approach the Government directly, Bill in hand and assurances that there was "unanimity amongst beekeepers" for compulsory powers. If it was not such a serious matter, his approach would be laughable. And why was he apparently so reluctant to collect meaningful statistics to support his claims? Could it be that he suspected that they would show him to be badly out of touch? There certainly would have been less kudos attached.

And how are his Irish outbursts to be interpreted? How did they really relate to the happenings at the time? I cannot believe that it won him any new friends and none of his existing associates expressed support for his views, but they seemed powerless to change them. What did he think it would achieve? If we are to believe W. Herrod-Hempsall, Cowan achieved an, albeit somewhat limited, personal victory over Rev. Digges regarding copyright (another chapter), but it was clear that Digges was not going to go away, and this attack was only likely to increase his resolve, and why did Cowan not use the legal case in his attack on Digges in the foul brood saga? .

Although he had retired to Somerset and rarely attended Council meetings, right up until his death Cowan insisted on being kept informed on bee legislation, He probably, correctly, considered it unfinished business, but I doubt if he felt that he was personally responsible for the failures. Writing his history of the B. B. K. A. he obviously revelled in, and felt very important dealing with members of Parliament and Government officials.

The inability to obtain legislation for the control of bee disease was a major failure for Cowan, and it is surprising that one of the offers to resign that he claims to have made, was not accepted. It reflects the perceived ability of the others.

Chapter VIII

Equipment

In the early years of his beekeeping, Cowan indulged in creating several new pieces of equipment, his engineering background proving useful. He always claimed that he did not demand royalties for anyone that wished to manufacture and sell any of his beekeeping inventions, just that they acknowledged that it was his design, and appended his name.

The Cowan Hive.
This double walled hive featured in every edition of the *British Beekeepers Guide Book,* even though it had been superseded by 1890.

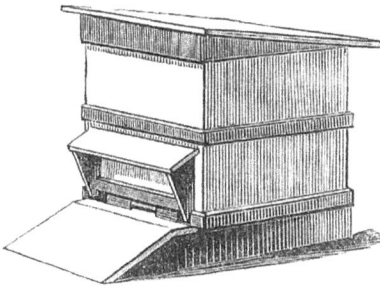

Perspective view of Cowan hive

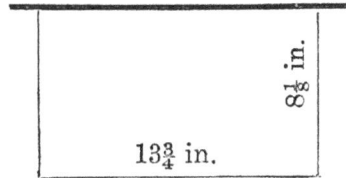

Frame dimensions

Before giving details of his own hive in the *Guide Book*, he described the general principles of hive design, which are as applicable today as they were 130 years ago.

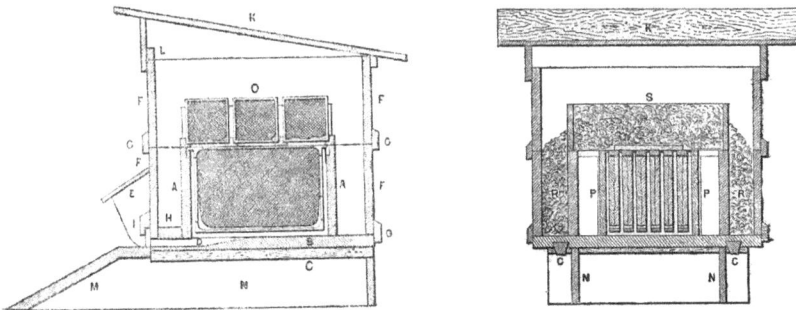

Sections through the Cowan, Summer and Winter arrangements

His hives were made to take either ten or thirteen frames, the smaller of the two in agreement with his statement that between 1800 and 2000 cubic inches was generally considered to be the best size for a hive in Britain. The dimensions were: -

Body box. 1" thick pine, 14 ½" deep, 8 7/8" high, both inside measurements. The width depended upon the number of frames it was to hold. Sides rabbeted with strip of tin to support the frames added. Later he claimed that the depth was chosen so that it could be obtained from a 9" plank after it was planed, but this was challenged.

Frame. He allowed "barely 3/8" between the inside of the body box and frame side for his hive, but in his hive design principles he wrote "It has been found in practice that if the frames are made so as to leave a full ¼" of an inch passage, round the ends the bees will leave this space ('bee space'), clear." He would later chastise those who gave vague quantities in their recipes etc., but it could be argued that 'barely 3/8' is very close to 'full ¼'.

The different dimensions given for 'bee space', leads to two possible sizes for his frame. Adding to the confusion, Herrod-Hempsall in *Beekeeping with Pen and Camera* gave Cowan's frame dimensions as 13 ¼" x 7 ½". Whichever dimension is used it is clear that the British Standard frame was firmly based on his model.

Floor board. 1¼" thick pine with a sloping passage gauged out to provide access for the bees.

Outer case. 5/8" deal (a cheaper timber than pine at the time), large enough to leave a dead air space around the body box, resting on floorboard as shown in diagrams.

The Summer diagram for the hive shows sections in place, but a super of the same size as the body box was also a possibility, either of the same height as the brood box or 5" high. The diagram for Winter shows the gap between inner and outer boxes, filled with chaff, supported above the body box by strips of wood, leaving a space above the frames for the bees to travel across. This was an arrangement that he reported upon having tested it on just one hive, over one winter. Insulation was considered to be a prime consideration by many bee-keepers at this time. He claimed that it was easy to strip down for cleaning, but removal of the chaff at the end of winter must have been a challenge. He also claimed that it was easy to move, an exaggeration because it was very heavy – a real engineering job, and consequently also relatively expensive.

Initially, he used Novice's metal frame corners to join the frames, but later abandoned them, moving to jointed corners, similar to those used now. He fixed foundation in the frame using a slot cut in the top bar. Cowan claimed to be the first to use the slotted top bar.

There is little doubt that this was a good and very successful hive, but he should have removed it from the *Guide Book* by the 20th century.

Strangely, I could find only one reference to Cowan's hive being exhibited, and that was at the Bath and West show of 1882, in Cardiff. In the class for best frame hive of substantial nature, maximum price 15/-, his hive was third, the same award as for his frame hive for cottagers, maximum 10/-. These prices were never achieved by manufacturers, and led Abbott to complain, on many occasions, that it was impossible for a manufacturer to make and sell some hives, for the prices claimed by their inventors at exhibitions, a remark not entirely directed at Cowan.

The Cowan hive did not feature in any contemporary publication, before 1881. However, W. Herrod-Hempsall in his book *Beekeeping New and Old,* published in 1930 and R. O. B. Manley in *Beekeeping in Britain,* 1948 stated that Cowan's hive was first described in the pages of *The English Mechanic,* in '1865 or 1866'. **This is incorrect.**

The English Mechanic and World of Science, first appeared in 1865, and did exactly as the title indicated. It comprised science-based articles, a subscriber exchange, for sale and wanted columns and a list of engineering contracts currently available. However, it was best known for the correspondence pages, where, as one contributor wrote "eccentrics and cranks capable of putting quill to parchment gave vent to their scientific hypotheses". It was not until April 1870 that the magazine announced that it was now to have space for amateur farming contributions. Initially there was only a 'trickle' of bee related items, but by September 1871 C. N. Abbott was a regular contributor and rapidly assumed the position as the magazine's 'bee man'. Many of Abbott's articles for the *English Mechanic* were repeated, in a slightly varied form in the early editions of the *B. B. J.* This goes some way to explaining how he managed such a prolific output in such a short time. Abbott's last entry was a few months before he launched the *B. B. J.,* after which, the bee content of the *English Mechanic* dwindled, eventually to nothing.

T. W. Cowan contributed many items on several different topics to the *English Mechanic,* but none on beekeeping, and certainly did not describe his hive in 1865/6. Additionally, there is evidence that he was not even a subscriber for the first five years that the magazine existed. His first entry is not until October 1869. He used the exchange and for sale columns extensively and I have given brief details of his entries in all categories in Appendix 1. They provide a small insight into his life before his major engagement with bee-keeping.

A search of the *Journal of Horticulture* for the same period produced an identical negative result. However, in his paper on wintering bees in October 1879, he stated that some of his hives were double walled. He did not say that they were of his own design, an unlikely oversight if that was the case. Why he even considered double walled hives to place in a stable loft is bewildering. A year earlier, Abbott in the *B. B. J.,* when answering a query on the size of the frame in the Cowan hive, wrote:

> The hive used by Mr Cowan is substantially the 'Woodbury hive;' the same as described in *Beekeeping for the Many,* published many years ago by the editors of the *Journal of Horticulture.*

There is no mention of a specifically designed and manufactured hive, single or double walled in this publication.

So why the need to lie about the entry in *English Mechanic,* and who was the initial perpetrator?

At the 1874 B. B. K. A. Crystal Palace show, Cheshire won the class for best hive. It featured in his publications and was being advertised for sale by James Lee by the end of that year. In his December 1876 "Abridged Price List of Apiarian Supplies for 1877", Lee advertised the - 'Cowan hive for 15/-, double cased and supers, 37/6.' The 15/- price was for basic hive – no lifts, supers etc. It did not feature on his list for the previous month. In Abbott's list in the February 1877 edition of the *B. B. J.* he included the Cowan hive at 15/-. Whilst the Abbott entry is unclear, there is little doubt that Lee's 37/6 hive is the one in the 1881 Guide. Could it have been that it was exhibited

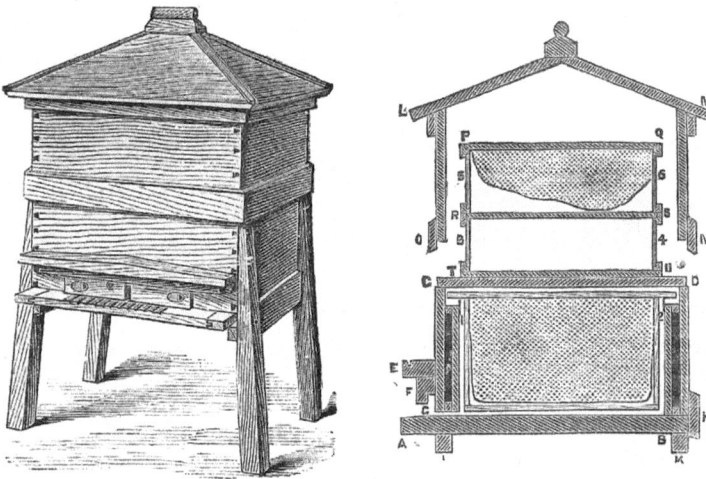

Cheshire's prize-winning hive

for the first time in 1876 but not a prize-winner? could it have been unplaced behind Cheshire in 1874? Unfortunately the Abbott reporting of the early shows was not exhaustive. One hive that Cowan did exhibited in 1876 was based upon the Giotto principle – a 'long' hive, but it did not feature again. So my somewhat insecure conclusion is that his hive was available in 1877, but details were not published until 1881.

As with the Cowan hive, the frames in the Cheshire version are Woodbury size. It was made with either an integrated or independent stand. Comparison of the Cowan and Cheshire hives shows many similarities. The Cheshire version predates that of Cowan by at least two years, and it is this that I believe has led to need for the *English Mechanic* misinformation.

W. Herrod-Hempsall wrote on several occasions that T. W. Cowan was the inventor of the first double-walled hive, something that I do not believe to be correct, and only made possible by the invention of the *English Mechanic* story. However, I think it highly unlikely that Herrod-Hempsall would have encountered *The English Mechanic,* and I doubt if it was normal reading matter for the other members of the 'London Club'; so, Cowan has to be the source. There are other dubious small episodes that involve the Herrod-Hempsall, Cowan and *The English Mechanic*, and these are mentioned elsewhere.

The Cowan Honey Extractors.

Although there were allegedly, reports of a machine in use in France earlier, the invention of the honey extractor is generally credited to Hruska in the mid 1860's. The Americans were the first to develop it into a marketable device, and those first used in Britain were imported at great cost. But it was not long before manufacturers in Britain were copying or modifying the American designs, and marketing their own efforts.

This was an area where Cowan had a major advantage over many others, benefiting from the knowledge gained when involved in Engineering manufacturing for fourteen years. His prototype machines were initially manufactured by a neighbouring business in Horsham. Herrod-Hempsall in *Beekeeping New and Old,* informs that Cowan:

> ..... commenced his experiments in 1873, and no less than
> thirteen different patterns of extractors were constructed,
> until by repeated trials and alterations, he evolved that
> which became universally popular (working upon the principle
> that America was not a part of the Universe at that time!),
> to the exclusion of all others. In experimenting, Mr Cowan
> extracted the honey from hundreds of combs, each being
> carefully weighed after the operation, and then given to

```
the bees for them to clean; when quite dry, they were
weighed again.  The difference in the weights enabled him
to determine the amount of lost honey. And by this means he
was able to arrive at the most suitable distance to fix the
cage from its axis.
```

The questions regarding the unscientific approach are many, but it was 1873, and this is W. Herrod-Hempsall reporting.

The first public mention of a Cowan honey extractor occurs in the report of the 1874 Chrystal Palace show. In the class for the best honey extractor there were only four entries, two from A. J. Starling of Tottenham Street, Kensall Road, London, who had been advertising them in the *B. B. J.* from December 1873, one by Mr Walton of Leamington and one by Cowan. Those of Starling and Walton were very similar tangential machines, Waltons had a "multiplying cog wheel bearing" whilst Starlings were operated via a lathe strap multiplying set up. The three machines drove rotating cages in a static cylinder.

Starling extractor   Walton extractor

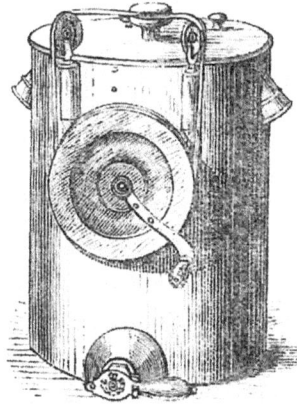

In C. N. Abbott's words:

```
Mr Cowan's machine was of a very different character; the
cylinder was similar, and it was mounted on a similar framework,
but  the  internal  economy  of  the  thing  was  differently
arranged.  Instead of the combs of honey standing at right
angles with the radii as a circle (as with the others), and
forming two sides of a square within the circle in which
they revolve, and which necessitates the reversal of their
positions, as each side of the comb is operated upon, they
stand perpendicular as radii from the centre of the machine
in the same plane with it's spindle; and the argument appears
```

to be, inasmuch as the cells point slightly upwards, towards the top bar of the frame of comb, if the frame is placed on end with it's top bar outermost, the cells will have their inclination outward, and at the angle which most facilitates the escape of the honey from them.

It was a radial machine, but not the first. It closely resembles one that Hamet described in his magazine *L'apiculteur,* of 1867, and he stated was in use in France as early as 1860, predating Hruska, but there is no evidence to support this.

Shortly after the Crystal Palace show, the *B. B. J.* carried the following advertisement:

HONEY EXTRACTORS. James Lee begs to announce that he has undertaken the Manufacture of HONEY EXTRACTORS on Mr Cowan's principle, one of which was exhibited at the Crystal Palace, and considered by many to be superior to any yet made. Price £2 17s 6d. Address – James Lee, Hive Manufacturer, Bagshot.

A year later James Lee had ceased to advertise this machine replacing it with another Cowan designed extractor.

Eleven years later Cowan reported upon an extractor by M. Buhne-Lauban, of Schlesien, which he had encountered whilst abroad: he said this machine used the same principle as his first extractor. The frames were held in horizontal cages, which enabled a shallower tank to be used, making it a table-top device, but it was never marketed. Cowan claimed that he successfully used his radial machine in his apiary, for many years, before donating it to the B. B. K. A. for their museum. He had found that when extracting new combs at high speed with it, they could be damaged, adding that it was not successful with those 'less experienced'. It was many years before the Americans exploited the advantage of the radial system for large machines.

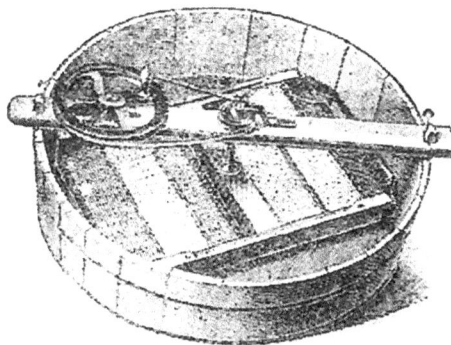

The Buhne-Lauban extractor

At the 1875 London show, Cowan entered three honey extractors. The first was his 'Cottagers', an example of his ability to think laterally, unfortunately, unlike his radial device, it does not produce a good machine, and it is surprising that he exhibited it with the other two. Of it Abbott wrote:

> It is a machine on the revolving-can principle, certainly an advantage as regards the prevention of undue circulation of air, and the consequent chilling of the brood, but as its motion is communicated from the bottom (the gearing being encased in the box upon which the machine stands), and it revolves upon an upright shaft, offering no means of steadyment by holding at the top, we fear it will not in use be quite as steady as desirable.

There is no further information available, so it is not possible to know how the motion was implanted to the drum. I was unable to discover any merchant offering these for sale, and it did not reappear at shows. A major use of the extractor at this time was it's ability to remove the honey from brood combs, leaving the sealed brood in place; chilling the brood was a major consideration and it was felt that this type of machine was a greater risk in that respect.

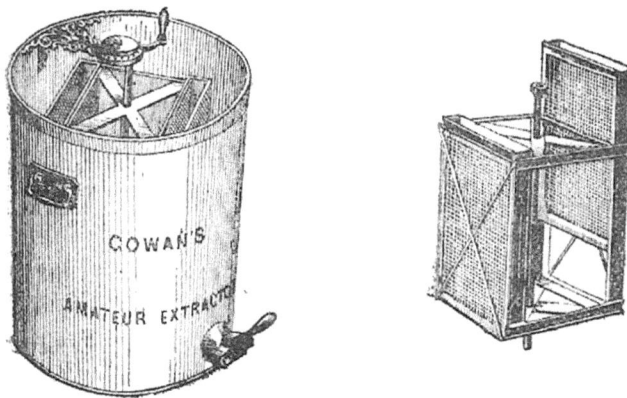

The Cowan 'Amateur' extractor showing internal cages

Cowan also showed a new machine the 'Amateur', which Abbott said had nothing special in its construction. It was included in the first edition of the *Guide Book* from which the engraving shown is taken. It is the simplest of tangential machines, capable of taking only two frames at any one time. The tinned iron can is 18" in diameter and 24" high with a conical bottom and treacle valve. The frames were held in two wire mesh baskets, as shown, on a framework joined to a central spindle, which is rotated by the handle at the top. When one side of the comb has been extracted, the cage is withdrawn from the framework supporting it, and reversed. It is, as he implied, an extractor for the small-scale beekeeper, who is not too concerned about the time that it will take. By 1881, G. Worldridge, of Station Road Horsham, a neighbour of Cowans, and W. W. Young, of 150 High Street, Perth, were both selling this extractor, at the same price of 30/-, "Easy and simple, a child can work them".

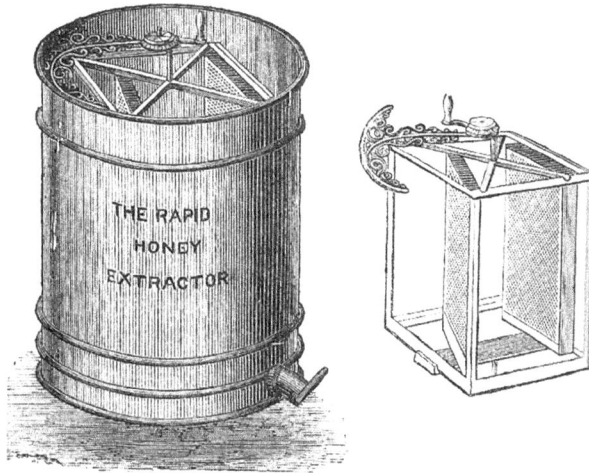

The Cowan 'Rapid' extractor

His third hive was 'The Rapid', and was a logical development of 'The Amateur', the arrangement not requiring the combs to be taken out for both sides of the comb to be extracted. As can be seen, it does not have a central spindle, leaving space for the two cages, which are pivoted at one corner, to be swung through 90 degrees after the first side of the comb has been extracted. The arrangement results in wasted space in the 'revolver', but the can was smaller than the 'Amateur', being 20" high and 17" in diameter. It was awarded the first prize. However, Abbott reported that 'Novice's' extractor from America was the neatest and liveliest machine at the show, with "peculiarly nice" drive mechanism, and "when set in motion it did not know when to stop". It had been in production for some time. However, it was considered to be too light for English wear, an extraordinary statement given the relative honey yields of the two Countries. Hunter had imported the machine "to give an impulse to English inventors" but it was beaten by an unfinished, just developed, device entered by the Chairman. It was foreign, inferior by definition. However, someone thought well enough of the mechanism to steal it! Although the

'Rapid' was priced for the show at £3, when it came onto the market both Worldridge and Young were offering them at 35/-.

Root in his autobiography mentions the incident, feeling somewhat aggrieved at the result. His mechanism was from an apple peeler, but Root made no secret of that.

Although Cowan claimed that he did not sell equipment, Abbott reported that all the honey extractors exhibited were demonstrated and sold.

The 'Rapid', with some minor improvements, was placed first again in the show at the Alexandra Palace of 1876, much to the annoyance of Abbott; a new machine by Cowan, the 'Express', was only highly commended and not described. The price quoted at the show for the 'Express' was 55/- but I do not think it was marketed. The 'Rapid' was the machine featured in the first eight editions of the B. B. K. A's booklet *Modern Beekeeping*.

In the 1879 show, Cowan appears to have exhibited only two extractors, the latest developments of the 'Cottager' and the 'Express'. In the 'Cottager' he had developed "an ingenious method of reversing the combs without having to remove them from the can", but we are spared a description. This was placed third. The 'Express' was place first and we were informed that it could take two frames, reverse them automatically by reversing the direction of the operating handle. Editor of the *A. B. J.*, T. G. Newman, who was in Britain at the time, purchased it to place in his 'museum'. It was normal for inventions to be donated; he wrote that automatic reversal of combs was "an entirely new thing in England but Mr. A. I. Root contrived a machine very similar to it in 1873."

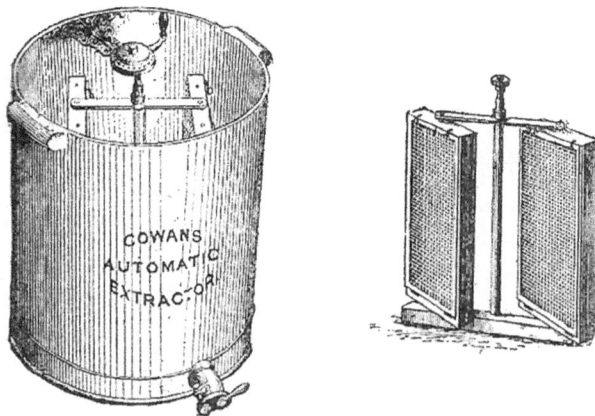

The Cowan Automatic extractor

At the 1880 show Cowan was first with his 'Automatic' a variation on the

'Express' and third with his 'Rapid'; a year later he was first and second with the same two. He appears to have settled for three machines, 'Amateur', 'Rapid', and 'Automatic', a logical range that covered most budgets and uses, prices being 30/-, 35/- and 50/-.

It was at this stage that all development ceased. His extractors described in the first edition of the *Guide Book* are the same as those in the last edition, although arranged a little differently. The cessation of honey extractor development appears to have coincided with his move away from South Street, Horsham, and Worldridge, who was responsible for the practical prototypes and probably assisted in the development.

Amos I. Root manufactured and sold variants of the Cowan machines in America. They were marketed for between $11.50 and $35.00 in 1900. £1 = $5 at the time.

A Group of the Cowan Family.

The sales of the Cowan Reversible Honey-Extractor are steadily increasing. Over 200 sold in July alone. You are sure to be pleased if you order one of these Extractors for we have no complaints of these machines. Be sure to specify Root's Cowan

For sale by all leading dealers and the manufacturers.

The A. I. Root Company,
MEDINA, OHIO.

Roots range of Cowan extractors

Tin Sections.
Cowan exhibited these at the Crystal Palace show of 1875, but without success. Cheshire described them thus:

> ...an admirable arrangement of boxes for receiving surplus honey. A number of these in tin, inverted side by side in close contact, make up the size of a large super, over which passes a good non-conducting cover. There is no

greater loss of heat than if all were in one. Each box contains a piece of wax sheet upon which the bees construct their comb. As completed, they are removed, and a lid put on, thus packing neatly and at once 3lb or 4lb of honey.

Unfortunately there is no accompanying engraving. They had already been overtaken by wooden sections imported from America, and were not exhibited again, or marketed, and might have been devised as an answer to the American product or could have been imported from there.

Glass Sections.

Yet another item that either Cowan claimed, or others claimed on his behalf, to have been the first to devise and exhibit, but without convincing evidence. Although there were claims by two individuals that they used sections with one or two sides of glass as early as1878, the first fully glass section appears to have been exhibited at the B. B. K. A. show in 1886. In response to a query in the 1888 *B. B. J.*, at least three individuals claimed ownership! They were the property of T. Bonner-Chambers who gave details of their construction but did not sell them. One of his letters carried the following editorial footnote:

We have received from our correspondent one of his glass sections. We can easily understand the admiration they excite; they are very sweet, cleanly, and attractive. We understand that a patent is about to be taken out for making glass sections; this we think is to be deprecated in an article which very probably may be in much request in the future; the more so as generally the words of a patent embrace much of the ingenuity of those that have preceded the patent.

For the total hypocrisy of this statement please refer to the chapter on Cowan's industrial life. The patentee was John Hewitt, which explains much. There was little difference between Hewitt's version and that of Bonner-Chambers.

In a later editorial, Cowan went into great detail of the disadvantages of glass sections, although accepting, that visually, they were very pleasing. He claimed that it would be difficult to make them for less than 1d each, (he quoted glass as costing 3d per square foot), and in the same paper a manufacturer was selling them at 3/- per dozen, (3d each). Later still, again to undermine Hewitt, he wrote "anyway a manufacturer is to market a different version shortly, so that will spoil the party". More in chapter on Hewitt.

Glass Separators.

Cowan alleged that he exhibited glass separators for section crates, at the Crystal Palace show of 1875, but unfortunately neither Abbott's report nor that in the *Journal of Horticulture* mentions it. He made this statement in response to Abbott reporting, in 1881, that Mr. J. Stewert, was the first to use separators in 1876. Cowan added that the sections that he was awarded first prize for at the Alexandra Palace show in 1876, were worked using them. He used them until the tin version was imported from America. Cowan also said that he had used glass between combs in supers. He ended his letter:

```
I am sure you will excuse my pointing this out, as you wish
to have a correct record.  I do not mean to say that I was
certainly the first to use them, but I believe I was.
```

Skeps.

At the Alexander Palace show of 1876, Cowan exhibited some "large and well made skeps with queen excluders etc", even though he was opposed to the continued use of the skep. The purpose was to demonstrate to 'cottagers' that if they used the larger flat-topped skep rather than the usual small domed type, it was possible to super it to obtain honey and avoid destruction of the bees. He had become resigned to the fact that 'cottagers' would not readily move from the skep to the bar-frame hive. There was nothing novel in this.

Wholesale feeding arrangement.

Whilst residing in South Street, Horsham, Cowan had thirty-six colonies, thirty kept in lofts above his stables, four in a bee house, and two in the open on stands. In 1875 he exhibited a system for wholesale feeding that he had used on his indoor hives whereby the food was conveyed by pipes, from a cistern, to the hives. Unfortunately the only information that I have been able to find is that the idea apparently suggested itself to him whilst observing cattle drinking when standing in their milking stalls. In my youth these devices consisted of a bowl with a paddle attached which the cow pushed with her rather large nose, opening a valve that released water into the bowl, for the animal to drink. I am intrigued to know how this was adapted for bee use.

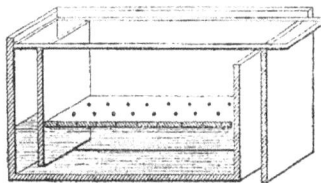

The Cowan half-gallon feeder

In his paper on wintering he described a different method that he referred to as 'bulk feeding'. In the autumn he selected a strong colony, supplied it with frames of empty comb and put a half-gallon feeder on it, using a paraffin stove to keep up a high temperature in the hive. When the empty combs had been filled and sealed, they were removed to other less strong colonies, and the exercise repeated. Neither of these experiments were mentioned again.

Hive Ventilating Equipment.

This piece of equipment was first mentioned by Cowan in the discussion following F. R. Cheshire's paper in early 1879 - *Abdominal distension of the Hive Bee during winter and the means of checking the same.* Cowan said "the idea being suggested by the Galton fire grate". The Galton fire grate took in air underneath it, from whence it passed into a warming compartment at the rear, where it was heated by the fire, after which it was then channelled back into the room through a grill at ceiling level.

The equipment was further described in Cowan's paper on "Wintering bees" at the October quarterly of the same year, with an accompanying engraving.

It was an attempt to continuously change the atmosphere in a hive without reducing the temperature or creating a draught. The tin box in the centre was 1" square and 4" long, but no dimensions were given for the tubes. He described its operation thus:

> This frame is placed in the centre of the hive, which is contracted in such a way that the bees are forced to cluster around the tin box. On the top a close fitting crown-board is placed, and the hive is raised one-eighth off the floor-board, and the entrance contracted to about half an inch. Now, the heat of the cluster warms the sides of the box, which at once causes the fresh air from the outside of hive to enter at the lower end of the tube, and in it's passage through the box the temperature is raised, and it flows out of the top tubes, and is diffused through the hive. The warm air thus entering at the top of the hive drives down the cold air and all the noxious gasses. A small tube at the side of the box, having a funnel reaching to within an inch of the floor-board, and having an outlet outside the hive at the top, also assists in getting rid of the gasses without creating a draught. In this way the hive is constantly supplied with pure air of an even temperature ….

He claimed it a success but had only tried it on one hive. It was included in the first and second editions of the *Guide Book,* but had been removed by the fourth, three years later. As usual there is no information from its inventor as to why he discontinued it, but as every beekeeper knows it was a daft idea in the first place. Little purpose would be served by listing reasons for it's uselessness, the disappointment has to be that someone who labelled himself an 'Engineer' and 'Scientist' invented it, and that no one was brave enough to point out the absurdity of it, even in 1876!

Non-alcoholic Mead.
No, there is nothing wrong with your spectacles. Cowan was a strict teetotaller, something that did not stop him giving views of the relative merits of the wines of the various places that he visited, or strongly defending the right of beekeepers to make mead.

The first mention of his quest to produce a honey based drink was by George Henderson, Librarian of the B. B. K. A. at the time, when he reported on a visit to Cowan at Comptons Lea in 1881. He wrote that his host had:

> ......said he was endeavouring to distil from honey some non-alcoholic drink, which if successful, he proposed to call 'Melidone'. ...... and I felt assured that a mind like Mr Cowan's, so versatile in resources and so fertile in inventiveness would be able to achieve the desired result.

As is often the case, there is no detailed information on the process or equipment involved.

However, it was not long before 'Melidone' made its public appearance. At the 1882 show at South Kensington in the class for mead, accompanied by it's recipe, Cowan was awarded first and second prizes, bronze medal and certificate. The recipe was not published, but in a letter criticising the judging at the show, a correspondent to the *B. B. J.* described it as "mechanically aerated

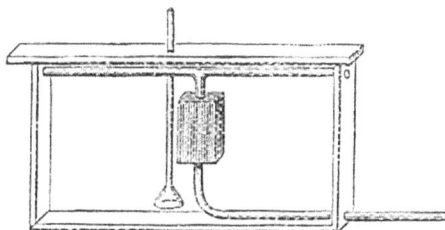

Frame ventilating device

water, sweetened with honey and flavoured with fruit juice." Not mead as we know it. The British Honey Company sold it, but I do not think it would have been a major seller, or a significant factor in the failure of the company.

Glass Jar.

At the 1882 show, Cowan exhibited a glass jar, and was awarded first prize for it. In the schedule for this class it was not made clear whether the exhibitor had to have manufactured the jar himself. In Cowan's case, there is some evidence that the jar he exhibited had been acquired on the continent, most likely at the Milan exhibition. A critical correspondent described the jar as "a horrible, heavy, green glass jar"; the apparent justification from the judges for placing it first was that "you can use them as a hammer to drive a nail with"; more importantly it was the Chairman's exhibit.

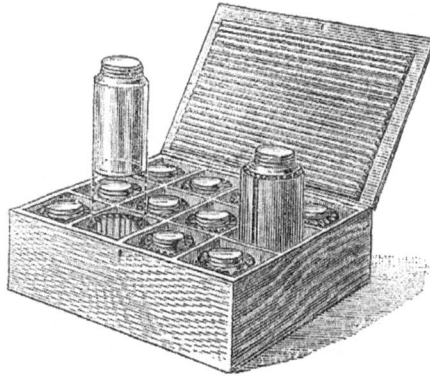

Crate for transporting honey in jars

Crate for Transporting Jars of Honey.

There was a new class introduced at the 1882 exhibition for crates to transport honey jars. Given the limited time that competitors had to prepare, it will come as no surprise that Mr. T. W. Cowan, who inserted the new class in the schedule, obtained first, second and third, plus silver and bronze medals for his crates to carry two, one, and half pound jars. They were essentially the same, just different sizes. The beekeeping public had to wait until the eleventh edition of the *Guide Book* in 1891, to view a sketch, shown here. It is difficult to see that it is different than boxes for transportation of any commodity in a glass jar at that time.

Crate for Transporting Sections.
Cowan exhibited these very early and they featured from the first edition of the
Guide Book. His first version is illustrated below and was a simple box, of such
a size to accommodate twelve sections, "glazed on two sides to insure careful
handling."

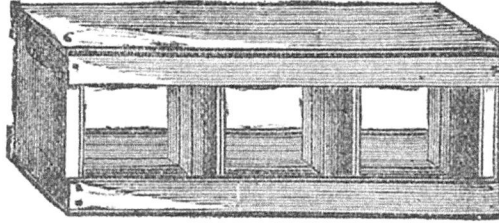

Section crate, first attempt

By 1889 the crate had undergone minor modifications and the engraving
improved to give a more professional job.

When filled with neatly-finished sections they have a very attractive
appearance, show off to the best advantage, and can thus be preserved for a
great length of time if kept in a warm room.

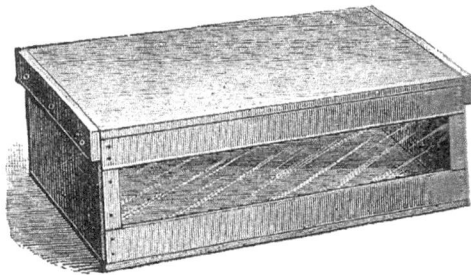

Section crate later version

Either from a desire to improve the article or as a result of a bad experience,
by 1891 it was further modified by adding springs above and below the sections:

The best marketing crates have springs top and bottom to
prevent the combs from being damaged by jarring during
travelling.

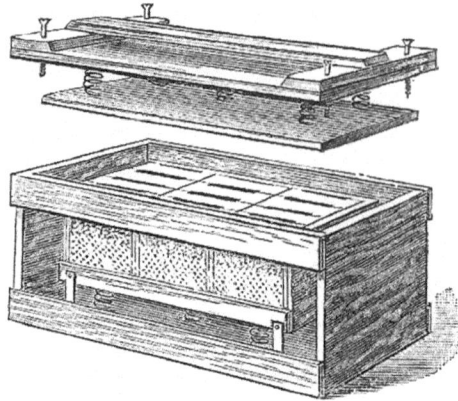

Improved travelling crate for sections

It remained in the *Guide Book* for every edition that followed.

Rapid Honey Evaporator.

T. W. Cowan first described this piece of equipment in the June 1883 issue of the *B. B. J.*, starting, as would be expected by justifying its existence:

> The honey harvest could be very much increased if as fast as the nectar is collected from the flowers and put into the cells by the bees it could be safely extracted. It is well known that after the bees store their honey and before it is sealed up, it has to undergo a process of ripening, or it would be liable to ferment. The heat of the hive assists in the process of evaporation, and only when the superfluous moisture has been extracted from it are the cells closed. Those who have had experience in extracting honey know the honey in the sealed combs is much thicker than that in the open cells, and that it is only safe to put the former into jars. There is a great deal of honey shown and sold that is unripe, but this in a very short time begins to ferment, and even becomes sour. The appearance of unripe honey is also peculiar. It has a decided green hue, and is not improperly called ` green honey.' In all books we are told not to extract from unsealed combs, and also for wintering we are recommended to extract all honey not sealed over, because the thin, watery honey is likely to produce disease. It is from its readiness to ferment that

disease (dysentery) is produced. When bees collect honey they put it into the empty cells, a little into each, so as to expose a large surface of the honey to the influence of the heat of the hive. If the income has not been very great during the day, the bees are able to evaporate the moisture sufficiently during a warm night to enable them to carry the honey from the lower cells to those above. As the honey becomes ripened it is sealed over, that at the top being ready first. If, on the other hand, the bees have collected a very large quantity of honey in the day, they are not able to evaporate it in the night, and, therefore, do not store it up above. All the cells being full, the bees returning with honey do not find anywhere to put it, and the consequence is that they waste their time in converting it into wax, and adding it to their cells. When bees are in this condition I think instinct (or reason) prompts them to make preparations for swarming. Queen-cells (which take a large quantity of wax) are constructed as a preliminary step. Now if we wish to prevent this we should extract the honey, and by extracting it daily a very much larger quantity of honey can be obtained than if we waited for it to be sealed over. We must also bear in mind that the sealing over is done at the expense of honey, twenty pounds being consumed to produce one pound of wax. Hitherto no satisfactory method has been devised for ripening honey, the ordinary cans doing very well when a small quantity of unripe honey is extracted with a large quantity of ripe honey, but they are quite unfit for large quantities. From experiments I have been carrying on 1 find that if honey is subjected to a heat under 200° Fahrenheit it is in no way injured either in colour or flavour.

Some of the above would be questionable today, and the temperature of 200 degrees F. (91 degrees C.), would definitely be considered excessive for honey - but is still below the temperature necessary to vaporise water.

Water is poured into A, unripe honey into B. The water passes into the boiler D and after heating a valve is opened and the hot water travels up through six shallow, interconnected tanks and back into A. When the honey is released it passes over the tanks in a zigzag manner, guided by vertical pieces of tin, attached to the tops, until it falls into C. The whole concept of the equipment is highly suspect.

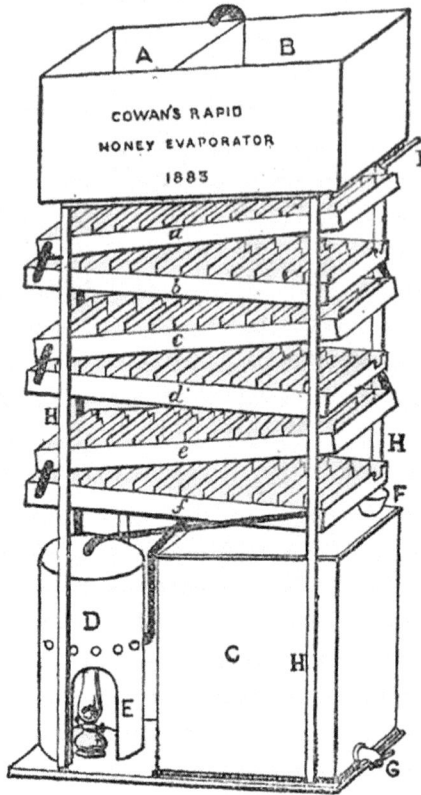

Rapid Honey evaporator

Cowan concluded saying:

In this way the honey travels a distance of 100 feet over a heated surface, and all the superfluous moisture is evaporated on its passage. If the honey is very thin, it may require to be passed through the machine a second time. The machine is constructed entirely of tin, as I find zinc or galvanised iron injures honey. From the rapidity with which the machine acts, I have called it ' The Rapid Honey Evaporator.'

In the *Guide Book* a chapter was introduced on Honey Ripening specifically to include the 'Rapid Honey Evaporator'; by the ninth edition in 1888, the diagram, and full description, had been replaced by a one paragraph description and by the eleventh edition in 1891, the whole chapter had been removed. The invention had a life span of about six years. It received a mention in Root's *A. B. C. of Bee Culture* but with the comment :

I doubt that such arrangements will ever be found cheaper and better than to let the bees manage it after their old-time fashion.

As would be expected, Hewitt was not complimentary, writing:

…..all very well to those who can afford it, but bees collect in day and ripen at night; so extract in morning. Ignores the fact that when sealed it is ripe, so an even more reliable system.

It was an unfortunate name to give the device, and he later referred to it as the Rapid Honey Ripener. He had probably used 'evaporator' to distance it from a similar device being marketed in America at the time.

Bee Quieters – Smokers.

At the 1875 Crystal Palace show, Cowan was first, of three, in class 25 - the best method of quieting bees. His exhibits are illustrated.

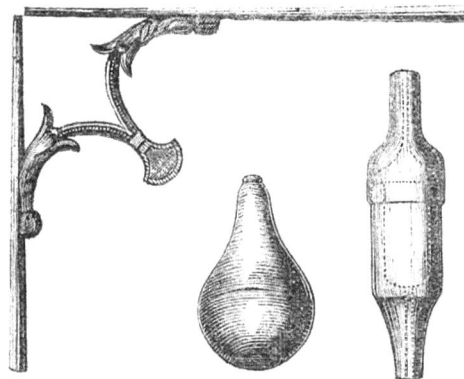

(b)   (c)   (a)

(a) is the smoker, the dotted lines indicating the inner construction. The lower part is filled with tobacco, lit with a 'vesuvian', and the top put back on. Blowing in the top will produce smoke from the bottom outlet. It can be held in the mouth continuously, leaving the hands free, and is suitable for non -smokers, like Cowan, to use.

(b) is a spray producer, after the pattern invented by Dr. Siegel in his steam spray inhaler. The perpendicular leg of the machine is placed in a small bottle of syrup, and the horizontal tube placed in the mouth, when by blowing, a jet of spray is produced by the air through the horizontal

```
tube rushing over the top of the perpendicular one, and
creating a vacuum in it into which the syrup rises, and is
blown to atoms.
```

This is typical Cowan, adapting a device to bee-keeping already in use for other purposes. He could have purloined his wife's scent spray, it would have worked just as well, which he acknowledged by the time he wrote the *Guide Book*.

(c) is an India rubber bulb, similar use to b. Cowan reported:

```
I generally give a few puffs of smoke with the smoker, then
remove the top of the hive, and next shower the syrup on
the bees with b; this instantly quiets them, and I then use
c to squirt some syrup between the combs.   After this the
smoker will keep them perfectly subdued.   I have succeeded
in uniting bees from two hives with this apparatus, as the
shower covers each bee with a fine spray of syrup, which, if
scented, is sufficient for the purpose.
```

In the same year, American Moses Quinby, devised a smoker with the bellows beside the fuel container, rather than at the end, as had been the custom before and two years later Bingham perfected it.

Foundation Fixing Aids.

Jig for fixing foundation

Devising jigs for mundane bee-keeping jobs, has been a pastime amongst those bee-keepers who believe that their colleagues are incapable of even the most elementary of tasks; most appear to be associated in one form or another, with fixing foundation into a frame. Cowan claimed to be the first to use the saw cut through the top bar of a frame, into which the foundation was fitted and held in place by the pinch pressure from the bar, often aided by the application of molten wax to the right angle between top bar and foundation. J. M. Hooker invented a spring clip that held the slit apart to insert the foundation, enabling a narrower slot to be used, and removing the need for the hot wax. Cowan first used foundation in 1864, two years after it was first imported into Britain. Initially the foundation was used in thin strips hung from the top bar.

For fixing full sheets of foundation, Cowan devised a jig. It was simply a piece of wood 3/8" thick, of width and breadth to fit inside a frame, with a couple of battens fixed along the back of it. The frame was laid upon the battens, a piece of foundation put upon the board and fixed to the frame top bar using molten wax, as before. He used a similar jig for fixing foundation into sections. Later he used a better method developed by Frank Cheshire.

At the 1882, South Kensington show, Cowan was awarded first and the silver medal in the class for inventions to advance beekeeping with a 'machine' for fixing foundation in sections. In a letter to the *B. B. J.* criticising the judging at the show, one correspondent wrote, "Cowan also won with an American foundation fixer of his own invention". Cowan never directly acknowledged that it was a copy but later in the *Guide Book* he called it "The Parker machine".

Solar wax extractor.
I have shown an illustration of Cowan's solar wax extractor elsewhere, and although he designed, had it made, and exhibited it in 1883, I do not believe he claimed originality for it.

Uncapping knife.
There is no evidence that he claimed design ownership of an uncapping knife, but Root sold one in the 1890s that carried his name.

Queen securing apparatus,
In the reports of the 1883 shows, it was stated that Cowan demonstrated driving using a device for securing the queen. Unfortunately we are not reliably informed of how this was achieved. An anonymous correspondent to the *Journal of Horticulture* somewhat cynically, said it was a piece of shaped queen excluding zinc. Without the details it is as good as any explanation.

Queen excluder.
He does not claim to have invented the zinc queen excluder, just to have 'improved' on the length of the slots. In his later bee-keeping life he reported hat he did not use a queen excluder.

In his writing Cowan described many pieces of equipment, sometimes credited to the inventor, but often not. Many times, Cowan returned from his travels with apparatus that he showed at meetings. Occasionally he would have something made that he had seen an overseas beekeeper use.

The items above are mostly those that I found he laid some kind of claim to; they show that he applied his inventive mind to beekeeping problems as he saw them, but not all illustrate the application of a logical or original thought process. He was not alone; most bee-keepers were, and still are devising equipment that they believe the be-keeping world has long had a need for. There is little doubt that bee-keeping benefited form Cowan's activities in this direction.

Chapter IX

The Digges Affair

In 1880 the B. B. K. A. sent C. P. Abbott and William Carr, with assistant A. Timberlake, to Ireland with the bee tent, to 'spread the word' about the 'new beekeeping'. It was a financial disaster. Bourdett-Coutts and Peel (Herrod-Hempsall also added Cowan), made up the substantial loss. There were very few attendees at any of the locations and the organisation of the tour, as might be anticipated, was a 'shambles'. Upon their return, Carr and Abbott reported that beekeeping in Ireland was in a poor state with little enthusiasm for change. Perhaps the Irish bee-keepers did not like the idea of the English telling them what to do. Cowan referred to this episode at every possible opportunity, and I believe it was the catalyst for his ongoing problems with the beekeeping establishment in Ireland. However, it needs to be viewed in the context of the politics of England and Ireland at the time.

Despite the tour failure, the Irish Beekeepers Association was formed in the following year. In 1887, Henry Chevenix started the leaflet *Notes and Hints* that was distributed free to all members of the Association. This was replaced in 1901 by the magazine, *The Irish Bee Journal*. Prior to the appearance of the *I. B. J.*, the *Beekeepers Record* had been offered to members of the I. B. K. A. at a reduced rate. After very few issues of the *I. B. J.*, Rev. J. G. Digges, Secretary and Treasurer of the I. B. K. A. took over as editor from T. B. O'Brian; (sometimes O'Brien or O'Bryan); and eventually became Proprietor.

Cowan claimed that Digges only became interested when the *I. B. J.* started to make money. Digges certainly stated that he did not subsidise the magazine and the I. B. K. A. paid for any advertisements they placed in the magazine. On many occasions Cowan claimed that, at times, he financially supported the *Record* and *B. B. J.* and for many years the B. B. K. A. did not pay for it's advertisements in his papers. The Irish magazine had a considerably lower circulation than its English contemporary, and carried much less 'paid for' advertisements, so both Digges and Cowan could not be correct. When subscribers to the *B. B. J.* started to complain about value for money, the Proprietor responded saying that it was not possible to produce a larger paper for a penny. The problem was that all the American bee journals were managing it, as were several weeklies in Britain at the time.

In 1911, Digges added the *Beekeepers Gazette* to the *I. B. J.* The 'Cowanites' claimed that it was the *I. B. J.* with a different colour label, which is untrue. In the first editorial for the *Gazette*, Digges wrote:

> The Gazette and Journal will not differ in every aspect.
> Articles, Reports, Correspondence, etc., having interest
> common to both countries, will appear in both publications,
> but such matters as are applicable to bee-keeping on one
> side of the Channel only, will be eliminated from the paper
> which is intended to circulate on the other side.

What affect the *I. B. J.* and *Beekeepers' Gazette,* (or the *Scottish Beekeeper* when it arrived) had, on the circulation of Cowan's magazines is unclear but obviously they did not improve it. He was not best pleased. However, his biggest problem was that the disenfranchised or non 'Cowanites' now had an outlet dedicated to bees, unlike the *Journal of Horticulture.* So alternative views and criticism could be freely expressed and sometimes encouraged, without the threat of a 'put-down' in the accompanying footnote.

The Irish Bee Journal.

The understated *I. B. J.* Banner

The *Beekeepers Gazette* Banner

Cowan would later write of the I. B. K. A. in relatively glowing terms whilst it was under the leadership of Henry Chevenix. But he seems to forget that there were several disagreements between the two. As an example: when, in 1888 Cowan wrote criticising the Irish hive of the time, Chevenix responded saying that they had not used a depth of 8 7/8" for the depth of the body box because a standard 9" board could not be planed to that size. He also pointed out that the Cowan sentence "….did not adopt the same size <u>frame</u> as the British standard" was untrue. It was a perfectly reasonable defence of the Irish hive and their right to have it. Cowan, of course, was not having it, repeated much of his original argument, saying that they found no problem with getting 8 7/8" from a 9" plank because they used quality timber and that there was a printers error, the phrase should have read "....did not adopt the same size as the British standard". So we are to believe that Cowan's printers were now adding words!

Also, at this time, in respect of Foul Brood Legislation, Ireland had acted independently, and apparently with greater success than England. For Cowan, the failure to get legislation in England to control Foul Brood was a major serial failure, which, in his eyes, should not be 'laid at his door'. The Irish, Digges, bee farmers, et al were likely to suffer the wrath of his pen, and be made responsible for his failure. The attempt to exclude Ireland from any legislation in 1904/5 was childish. This episode in English/Irish beekeeping relations is fully covered elsewhere.

The Irish were not happy with the W. B. C. hive. By the turn of the century, they had developed their own hives, the Congested Districts Board and the Federation, but continued to use the British standard frame size even though they would have preferred to have had their own. This, as with their initial hive was a pragmatic approach, enabling them to easily and inexpensively obtain foundation.

With relations between the two Associations and their leaders simmering nicely, in 1904, Digges published *The Irish Bee Guide, a Manual of Modern*

THE

IRISH BEE GUIDE,

A Manual of Modern Bee-Keeping.

BY

REV. J. G. DIGGES, M.A.,

Honorary and Member of the Examining Board, Irish Bee-Keepers' Association;
Editor, Irish Bee Journal.

PUBLISHED BY
IRISH BEE JOURNAL Office, Lough Ryan, Co. Leitrim, Ireland.
EASON & SON, LTD., Dublin and Belfast.
SIMPKIN, MARSHALL, HAMILTON, KENT & CO., LTD., London.

[Copyright. All rights reserved.]

Frontispiece and title page, *Irish Bee Guide*, 1904

Beekeeping. The Digges/Cowan parallel was now complete. Not only had the B. B. K. A. lost members to the I. B. K. A., the *B. B. J.* and *Record* had lost subscribers to the *I. B. J.*, but now Cowan's *Guide Book* had opposition from Digges's new book, and the use of 'modern' in the title must have really annoyed Cowan. Digges effort was considered by many to be better, particularly in the subject arrangement and writing style, and it had the advantage of coming later. Although it contained many photographs, some of the line drawings were poor. It was just another textbook, instructing on practical beekeeping.

Unfortunately, the only account of what allegedly transpired is provided by Cowan, who supplied the information to the Herrod-Hempsall brothers prior to his death, with instructions that it was not to be made public by whoever was editor of the *B. B. J.* at the time, until after both he and Digges had died. The truth has therefore been subjected to two filters or distortions. This illustrates the control that Cowan had and still correctly believed, that he could exercise even after death. Digges died in 1933, seven years after Cowan.

It was not until 1936, when the dynasty was under severe attack from many directions, that the Herrod-Hempsalls published the details of the 'Digges case'. It appeared in an attempt to quell the outrage being expressed following a series of articles that appeared in the *B. B. J.* during 1935/6 under the title *The B. B. K. A. and its Critics.* Joseph Herrod-Hempsall was the editor at the time, and it was published as an editorial, but there are strong indications that it was the work of his brother. At the time William was completing the second volume of *Bee-keeping New and Old* whilst staying at Buckfast Abbey, allegedly recovering from mental problems. Neither of the Herrod-Hempsall brothers was likely to let their mentor down, after all, they owed their position in the beekeeping establishment and their wealth to him. They had been carefully selected and groomed, although strangely, from a different layer of society, but as close to clones of their master as it was possible to get. As reported the incident went thus:

Cowan claimed that large parts of *The Irish Bee Guide* had been 'lifted' from his *British Bee-keepers Guide Book* and *The Honey Bee.* When initially approached by Cowan, Digges denied the allegation. However, following considerable correspondence, he agreed to meet with Cowan in London, both accompanied by Counsel, in an attempt to avoid court action. The case would have been heard in an English Court, where Digges would have stood little chance.

At the meeting Digges again denied the allegation. Cowan claimed that 13 of the illustrations in Digges book were copied from his work, and asked Digges to produce his originals of these, which he reluctantly did. They all had printed text on the reverse side, the implication being, that they had simply been cut out of Cowan's books, and supplied to the engraver. Why was Digges carrying such incriminating evidence and prepared to produce it? Why not stall or offer the printer's blocks? On the evidence of his writings he was not a fool. It does not appear credible. Cowan also claimed that Digges had 77 pages containing text copied from the *Guide Book.*

Searches have failed to produce the evidence from the meeting, but my understanding is that all that Digges had on this matter, upon his death, was passed to R. O. B. Manley, but went no further. How much Cowan gave to the Herrod-Hempsalls is unknown, but it would certainly have been carefully screened. However, if reports are to be believed, no private bee-keeping matter survived the twenty-four hours following William's death.

The preface to *The Irish Bee Guide,* page vii, written by Digges on 23rd May 1904, contains the justification for the book and his aspirations for it. Page viii carries a note detailing the arrangement of the book – each paragraph was numbered. He acknowledged 12 bee anatomy illustrations from Upcott Gill, (these were Cheshires); 20 from various others, and stated 81 were from original photographs by Digges and "pen and ink sketches drawn specially for this work by E. D."; E. D. was his brother. The number of drawings and photographs attributed with a figure number totalled 113. There were several photographs and diagrams that had been given no figure number, including 23 full-page plates. Cowan is not mentioned.

The 18th edition of Cowan's *British Beekeepers Guide Book* appeared in 1903 and was essentially the same as the previous two in 1900 and 1901. It had undergone considerable revision in the 14th and 15th editions of 1896 and 1898 respectively and contained 90 numbered figures; some of the more recently added photographs not being numbered, and were the only illustrations to be credited.

The 19th edition followed in 1907 and, as usual, heralded in the *British Bee Journal* - out "after considerable but unavoidable delay. 2000 already ordered". The following is from the Preface:

> So many of the illustrations as also descriptions have been copied from the 'Guide Book' by some writers, who have abstained from mentioning the source from whence they were derived that I have considered it necessary to rewrite a large portion of the work and have new illustrations made.

He acknowledged many of the photographs, and "to Mr. L. Upcott Gill who generously gave me a free hand to use illustrations from Mr. Cheshire's well-known work on bees, I am indebted for....."., accepting that he had used them previously without acknowledgement. He also acknowledged Sladen for illustrations showing queen rearing, a new section in the book.

1910 saw the re-publication of Digges manual, renamed *The Practical Bee Guide*. In the Preface he writes:

> The number of illustrations has also been increased by the insertion of 20 new blocks, while, of those in the first edition, I have removed 53, supplying their places with others, more accurate, deeming it wise, if not, indeed, necessary, to rely on my pen and camera for the illustration of manipulation, appliances, etc..

He acknowledges Cowan's works, but also a long list of 41 others headed by Cheshire, who gets a very special mention; making a point again?

In the note Digges states that of the 142 illustrations in the book, 120 are from original photographs by the author, and pen and ink sketches were drawn especially for the work. He then acknowledges 20 from other people, those from manufacturers and merchants being amongst those removed; and the remaining two? Whilst there is no evidence, it is very likely that Cowan had threatened to ban those commercial interests who supplied illustrations to the *Practical Bee Guide* from his own magazines if they continued to allow Digges to use their blocks: it would not have been the first time that he had used this form of blackmail! There are 118 numbered figures.

Logically the 13 illustrations objected to have to be in the initial line diagrams contained within the number 81 of the 1904 *Irish Bee Guide,* and the 53 removed from the first edition of the *Practical Bee Guide*. From my lengthy investigations I offer a sample:

(i). Illustrations of the queen, worker and drone bee are identical in *The Irish Bee Guide* and the 1st - 18th and 19th editions of Cowan's *Guide Book*. In *The Practical Bee Guide* of 1910 they were replaced with a photograph. However, the illustrations in Cowan's book are identical to those on the title page of the second edition of *The Honey Bee* by Edward Bevan, published in 1838. The size is the same and the positioning and markings of the wings identical, but they are not the same as those of Girdwoyn as believed by some. However, a queen bee, is a queen bee is......

(ii). Illustrations of a double rabbet joint are identical in *The Irish Bee Guide* and the 18th and 19th editions of Cowan's *Guide Book*. In *The Practical Bee Guide* of 1910 it was replaced with one of a chamfered inner wall.

(iii). Illustrations of the Porter bee escape are identical in *The Irish Bee Guide* and the 18th and 19th editions of Cowan's *Guide Book:* in *The Practical Bee Guide* of 1910 it was replaced with a much inferior effort.

(iv). Illustrations of comb foundation are identical in *The Irish Bee Guide* and the 1st - 18th editions of Cowan's *Guide Book*. Additionally, Digges had included two engravings showing the correct and incorrect orientation of foundation in the frame and, although the shading etc. is consistent with the first one, the engravings are inferior. Cheshire included two such plates in his *Bees and Beekeeping* Vol. II. In his 19th edition Cowan removed the original engraving of foundation and included, with acknowledgement, the two from Cheshire. In the *Practical Bee Guide* of 1910, Digges replaced all his foundation engravings.

There is no doubt that all four examples above the engravings were from the same block, lending credibility to the belief expressed by Annie Betts and R. O. B. Manley that Digges used them thinking that he had the owners permission. He did not however, include them in his acknowledgements, and with Cowan continuing to use the first three after the legal settlement without acknowledgement, the implication is that he was the owner. Bevan was not alive to argue otherwise. In the fourth case, it would appear that

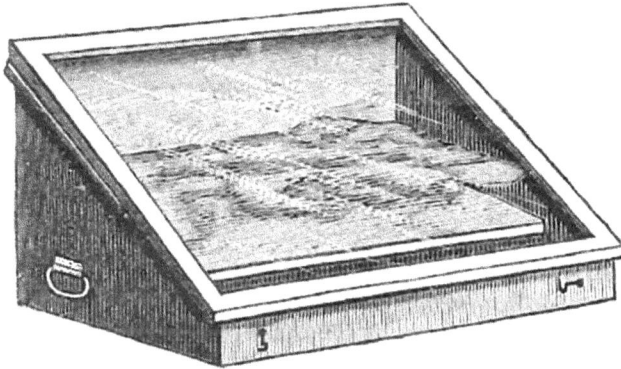

Cowan solar wax extractor

Digges solar wax extractor

Cowan had used the block, or copied the drawing from someone else without acknowledging such. This could have been brought to his notice during the legal negotiations, and he chose not to use it in the next edition of the guide in order that it was not obvious that he had used it for the previous 18 editions without acknowledgement. The same could be true of the Upcott Gill/Cheshire engravings. There is substantial evidence that this block was one of those involved in the original dispute over ownership of certain drawings concerning Cheshire, Abbott, Cowan and the B. B. K. A. in the late 1870's.

Herrod-Hempsall, in his report of this episode, did not detail any of the above, stating that the real damming evidence was the diagram of the solar wax extractor. He stated that although Digges had made some cosmetic changes, Cowan pointed out that he had repeated an error contained in the original, i. e. the metal floor had mistakenly been shown torn in Cowan's' book and is "faithfully reproduced" in *The Irish Bee Guide*. Reference to the two engravings below, whilst leaving little doubt in my mind, that one was used as the basis for the other, the reasoning of Herrod-Hempsall /Cowan is flawed. There is no tear shown in the metal floor of either, what is being referred to is the wax resting on the metal floor.

To my certain knowledge the solar wax extractor was ever present from the ninth edition of 1888 of Cowan's book. So Herrod-Hempsall would have us to believe that Cowan admitted to carrying a mistake in his *Guide Book* for 10 editions and 14 years!

Digges had also given details of the solar wax extractor of M. H. Read, complete with measurements. In *The Practical Bee Guide* of 1910, the diagram was replaced by a photograph, as it was in the next edition of *The British Beekeepers Guide Book*. But how different can a solar wax extractor look, anyway? Mine looks remarkably similar.

Also, in the *B. B. J.* article it is stated "…in the reproduction of his honey extractor could be plainly seen the scar by obliterating his name". Cowan always illustrated his extractors with his name and/or name of extractor on the can. Digges, in the *Irish Bee Guide*, only shows one extractor. The most recent Cowan *Guide Book*, 1903, showed cut-away diagrams of the 'Amateur' and the 'Rapid'. The last time Cowan published a diagram of one of his extractors in the simple manner to that of Digges, was in the 13th edition of his *Guide Book*, 1894. Both are reproduced for comparison. Herrod-Hempsall is implying that the same block was used with the text removed. This would have resulted in a black mark on the image in *The Irish Bee Guide*. There is no such mark on the engraving but it might just have been that he clumsily expressed it. Again, how different could an extractor be made to look?

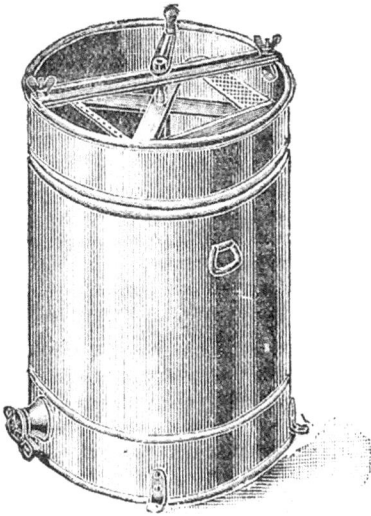

Digges illustration   Cowan illustration

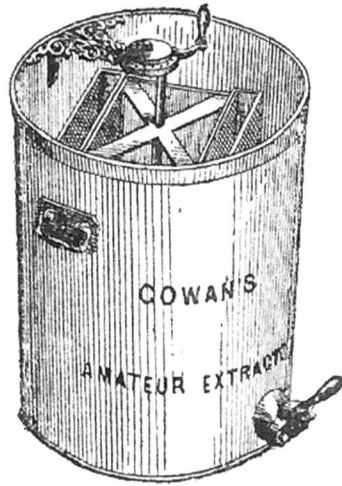

Digges also included in the *Irish Bee Guide*, a photograph of "extracting honey" with an extractor in the foreground that has a black mark across it. Could this be what Herrod-Hempsall refers? If so, why did Digges repeat the photograph in the first edition of *The Practical Bee Guide* with the mark still present, and not bring a response from Cowan? Also in that book, the engraving of the extractor is replaced with a photograph, which is obviously the same piece of equipment as in the previously referred to photograph and engraving. None of them look the same as any of Cowan's extractors without the mark; showing that the problem lay with the photograph or its reproduction.

However, Cowan also produced further evidence by annotating Digges book to show that 77 pages contained material "purloined from the writings of Mr. Cowan", and the remainder from other authors. Again, without the actual evidence to consult, it was necessary to carry out some form of forensic investigation. Fortunately, we now have the ability to digitally scan, compare and search; technology providing much needed assistance. I first compared *The Irish Bee Guide* with the *Practical Bee Guide* of 1910 believing that Digges would not have kept the passages that he had been accused of "purloining". I eliminated the extra sections that he had added, leaving those pieces that most likely comprised the contested passages. I then searched Cowan's 18th edition

Extracting honey' from *Irish Bee Guide*.

of 1903 for identical passages; obviously the smaller the match asked for, the greater the number of results. On the basis that Cowan had rewritten the parts that he considered had been compromised, I also compared the 18th (1903) and 19th (1907) editions of his *Guide*.

I failed to find large identical passages, but as would be expected when comparing two manuals, there were many similar entries. For example:

Irish Bee Guide. When the feeder is being removed for refilling, a piece of tin is inserted below it to prevent the bees escaping.

British Beekeepers Guide Book, 18th edition. A piece of tin is inserted below the mouth of the bottle when removing the latter for refilling to prevent the bees from escaping.

This appeared in the 19th edition unchanged.

Irish Bee Guide. Although in a favourable season an early first cast may be made profitable, after swarms should be prevented because they weaken the parent stock, and often render it quite useless.

British Beekeepers Guide Book, 18th edition. Second swarms, if early, may be made profitable [as separate stocks] but subsequent ones are (usually) of little value, even in favourable seasons and should be prevented for they so weaken the stock as to render it practically useless.

The square bracketed words were the "re-write" for the 19th edition.

Irish Bee Guide. When only a small quantity has to be dealt with the combs may be broken up and placed in a cheese cloth, or a fine sieve, over a pan of water all being then put into the oven. The wax will melt and drop into the water and can be removed in a cake after cooling.

British Beekeepers Guide Book, 18th edition. Where only a small quantity of wax has to be melted the comb may be placed in a fine sieve over a pan of water and put in the oven. The heat of the oven melts the wax, which drops into the water, and it can be taken off in a cake when cold.

Despite Cowan's claim that theft of his material had forced him to re-write much, there is only small change between the 18th and 19th editions.

There were other examples, but as with those above they were not exact copies. There could be little doubt that Digges had a case to answer. The problem was, not that he had used Cowan's work but that he had failed to acknowledge it. This was probably intentional, because on page viii of the book in question, he openly acknowledges many others, and states that the photographs were his, and pen and ink sketches were drawn especially for this work by "E. D." (his brother). However the evidence indicates that "E. D." used some of Cowan's illustrations, at least as a guide. The weakness of the prosecution case is that Digges's book was just another manual informing beekeepers on all practical aspects of the craft and inevitably contained the same information calling for similar illustrations and descriptions attached to them. For example, how different can an author illustrate, or describe the way to handle a frame when inspecting it from those preceding him?

Herrod-Hempsall, as usual in such reporting over stated the case and used highly inflammatory language. Digges was guilty as charged, but how Herrod-Hempsall thought that making this information public, especially with such language, would help him and the B. B. K. A. survive the onslaught that they were under at that time is impossible to comprehend. I also question what Cowan had passed to the brothers, and whether he actually gave them permission to use it. We only have the Herrod-Hempsalls' word for this. There was no mention of this from 1904 to 1936.

Digges conceded that he had transgressed. The negotiations that followed are reported in the *B. B. J.* article as Digges pleading for clemency and Cowan being the benevolent, generous man that we all know he was, or certainly as he and the Herrod-Hempsalls wish him to be remembered. There is no documentation available to support much of what was written in the *B. B. J.*, and it is always difficult to be convinced of a report that contains ver batim statements of what individuals said at a particular juncture, as well as including such words as "outrageous, dastardly, cunningly altered, untruthful and malicious," on the one hand, and "magnanimous" on the other.

Herrod-Hempsall informs us, that as an opening gambit, Cowan called for all the copies of the *Irish Bee Guide* to be handed over for destruction, including those already sold or in the hands of agents. Even in the early years of the 20th century it is difficult to see this being a practical proposition, the questions are many, but it would certainly have made *The Irish Bee Guide* a real collectors' book! If this is accurately reported, it shows a certain naivety on Cowan's part, but it is more likely again, to be a piece of Herrod-Hempsall's invention.

Herrod-Hempsall claimed that Cowan said:

> Digges pleaded so hard that that it would be seriously crippling him, that I consented to some compensation being paid instead.

The poor composition of this sentence points again to a Herrod-Hempsall effort. We are informed that the initial print run of the *I. B. J.* was 1,000, but there is no information on how it was financed, or the proportion in private ownership and hence, unlikely to be recovered. The *Irish Bee Guide* was published at 3/- by the Irish Bee Journal, Eason & Sons Ltd., and Simpkin, Marshall, Hamilton, Kent and Co. Ltd.

The initial proposed level of compensation, suggested by Cowan's counsel, was £50:

> ....but Digges pleaded so hard (that expression again), that I said I did not want to press him hard, but thought he ought to suffer to some extent, so I willingly agreed to let him off with £25 compensation.

The initial amount of compensation was, presumably, that requested by Cowan. Here it is being suggested that this was the proper amount, rather than the opening bid in a negotiation. Digges also paid £15. 15s. legal costs.

The accurate information is contained in the document signed by Digges and reproduced in the *B. B. J.* of October 1st, 1936. Manley claimed that the original was framed and hung on the wall of the magazine offices for many years.

> To
> T W Cowan Esq.
> 10 Buckingham Street, Strand, W.C.

In consideration of your forbearing to take proceedings against me in respect of the infringement by me of your copyright works "The Honey Bee" and "The British Beekeepers Guide Book" some of the illustrations contained in which I have used in my book "The Irish Bee Guide". I hereby undertake and agree as follows: -

To forthwith write an apology in the form agreed between us signed by our respective solicitors and hereunto annexed and transmit the same to you for insertion in the next issue of the "British Bee Journal".

To insert or cause the said apology to be inserted in the next issue of "The Irish Bee Journal".

To insert in all copies of "the Irish Bee Guide" at present unsold, opposite the acknowledgements which already appears in the book, a slip in the form agreed between us signed by our respective solicitors and hereunto annexed. To obtain from the publishers of the book and furnish to you a certificate that all the copies of the book at present unsold contain such slip and to forthwith cause all copies of the book at present in the hands of Agents to be returned to the publisher to be exchanged for copies containing the said slip.

That I will at once discontinue or cause to be discontinued the use of letters and extracts from letters as parts of advertisements of my said book in the "Irish Bee Journal".

That I will on the signing hereof pay to you the sum of £25 by way of compensation for the aforesaid infringement of your copyright works together with the sum of £15 – 15 – 0 the amount of solicitors costs incurred by you in the matter of the said infringement.
Dated the 26th day of October 1904.
J. G. Digges

Herrod-Hempsall gives the apology as initially demanded by Cowan, thus:

I desire to express my regret that in the compilation of my book, 'The Irish Bee Guide', I made use of the information and teaching contained in Mr Thomas William Cowan's works, namely, 'The Honey Bee' and 'The British Bee-keepers Guide

Book', without any acknowledgement; also that I have used
some of the illustrations from those copyright works without
permission, and in consideration of Mr Cowan forebearing
to take proceedings against me in connection with this
infringement of his copyright, I now apologise for so doing
and undertake not to do it again.

The version that finally appeared is given below as Cowan agreed it to be
amended because "he pleaded for his face to be saved":

(To the Editors "The British Bee Journal")
The Irish Bee Guide
Ammende Honorable.
Dear Sirs,
I desire to express my regret that in the preface to my
book, 'The Irish Bee Guide,' I did not, as I should have
done, express my acknowledgement to Mr T W Cowan for the
information and teaching contained in his standard works,
The Honey Bee and *The British Beekeepers Guide Book*, to
which I owe, in a large measure what I know of bees and
beekeeping. I also desire to express my regret that I have
(under a misconception of my right to reproduce them) used
certain illustrations from these copyright works without
first obtaining his permission. I need hardly say that I
deeply regret having infringed Mr Cowan's rights, and that
he may rely upon it that, now I am better informed as to the
law of the matter, I shall take every care that no further
infringement occurs, and I acknowledge the generosity of
Mr Cowan's behaviour in overlooking my error,
                          Yours Truly,
                    J. G. Digges

Yet again Herrod-Hempsall views this as an indication of the benevolent
nature of his benefactor, but it is difficult to see how it 'saved his face'. As
stated, this was a statement agreed between the two lawyers, the legal hand
being evident in the published version.

Ammende Honourable – public apology and reparation made to satisfy the
honour of the person wronged.

However, Herrod-Hempsall reports that Cowan considered that he headed
it 'Ammende Honourable' rather than 'Apology' because he did not "want to
humiliate Digges". Cowan continued:

Digges evaded the spirit of the agreement by copying the
title and inserting it in small letters instead of the

ordinary heading in 'The Irish Bee Guide'. Moreover, he
tried to save his face by printing it as a letter to the
Editor of the 'Irish Bee Journal', i. e. himself.

It is not known when Cowan wrote this, but it appears that in retrospect he was not happy with the agreement reached, but this is very strange reasoning.

It is an extraordinary apology for anyone to sign. The only legal dispute concerned Digges copying 13 of Cowan's diagrams, and reproducing some pieces of text without acknowledgement; I have shown that there is some truth in these accusations. However, as well as getting Digges to apologise for using some of his illustrations without acknowledgement, Cowan manages to persuade his opponent to concede that most of his beekeeping knowledge had been learnt from Cowan's two standard works. It is yet another example of the deification, (a Manley expression), of Cowan, and completely ignores the many other books on the subject that Digges no doubt studied and used. No doubt it satisfied Cowan's vanity.

If the only text books to be published were those containing original material, there would be very few, and of minute proportions. How many present day practical bee books give credit to authors of the previous decade, not to mention such classical writers on bees as Columella from several centuries ago. In the preface to the first edition of his *Guide Book*, Cowan, wrote that:

......most of the illustrations have been drawn on blocks
expressly for this work.........I have now to thank the many
friends and correspondents who have rendered me assistance
and advice, more particularly Mr. Abbott, who generously
placed at my service the wood engravings which have appeared
in the *British Bee Journal*, and Messrs. Cheshire, Hooker,
and Neighbour, for the use of the blocks illustrating their
inventions.

Although this does not give credit to those from which he first learnt his bee-keeping, it would have satisfied most as he obviously thought at the time. There was very little original content in the *Guide Book* and even less in *The Honey Bee*.

The 'slip' inserted in the *Irish Bee Journal* Reads:

I desire to express my indebtedness to the works of Mr T.
W. Cowan, viz. "*The Honey Bee*" *and* "*The British Beekeepers
Guide Book*" for the information and teaching contained in
those works. I also desire to acknowledge my indebtedness
to those works for some of the illustrations which I used, in
the first instance, without permission, but to the continued

use of which Mr Cowan has now kindly assented.

I have copies of *The Irish Bee Guide* with this 'slip' inserted, in the position stated, both full page size and smaller, at the back of the book and without it. The latter is presumably from the first run of 1,000 copies, and the others from later issues, and clearly shows that Digges complied with that part of the agreement, at least in three cases. Herrod-Hempsall maintained that the certification was never provided, nor copies in the hands of the publisher returned, to have the insert added. How would anyone other than the publisher know this? He was not part of the agreement. Digges book was published by: -

Irish Bee Journal Office, Co. Leitrim, Ireland

Eason and Son Dublin and Belfast

Simpkin, Marshall, Hamilton, Kent & Co. Ltd. London

It would have been easy for Digges to put the insert in those copies that he held for sale.

The fourth condition has little direct connection to this particular case, but Cowan was very much against, what was called "puffing letters" appearing in advertisements for books. The letters praising the *Irish Bee Guide* that I have seen were published originally in the correspondence column of the *Irish Bee Journal,* and these are very similar to those that appeared in the *B. B. J.* in respect of the *Guide Book.* Additionally, answers to queries were often responded to by reference to the relevant passage in Cowan's book, and every "Homes of the Honey Bee" seems to have required reference to the *Guide Book* as the sole source of that beekeepers' knowledge. If there is a difference it is across a very fine line. Herrod-Hempsall writes that Cowan wanted these letters banned because they led people to think that the book was Digges' own original work, whilst he considered the book to be a compilation. An accusation that could equally be made against both *The Honey Bee* and *Guide Book.* It is a very weak point and one that Digges chose to ignore without consequence. Herrod-Hempsall also points out that letters from people long dead were used to "puff" the book, but presumably they were alive when they wrote them - if indeed they actually existed!

In 1910, Digges republished his book, with a new title *The Practical Bee Guide,* without the copies of Cowan's diagrams and only minor adjustments to the text. There were two full pages of "puffing", one from the press the other from members of the craft and more in the preface. In the preface he acknowledged Cowan's two books, but then included nearly every author on bees in the English language. The work ran to 16 editions, the last in 1950. From the 8th edition (1936), R. O. B. Manley was the reviser of what little change there was. The preface to the second edition with its reference to Cowan was retained until the fifth edition of 1921. It was removed in the sixth edition published in 1928 two years after Cowan died, when Manley first became involved. There is some confusion over the early print runs of Digges book; one source quotes that initially 1,000 were printed, but the title page of the second edition states, 'ninth thousand', being the total print run of first and second editions. The third edition proclaims 'thirteenth thousand', and then lists 5,000, 5,000, and 3,000 for editions one, two and three respectively, which at last correlates. Mathematics not a strong area for the Rev. Digges? The manual eventually went to sixteen editions and seventy-six thousand copies, easily matching those of Cowan's *Guide Book*.

William Herrod-Hempsall followed in his master's footsteps writing *Anatomy, Physiology and Natural History of the Honey Bee*, and *The Bee-keeper's Guide to the Management of Bees in Moveable Comb Hives,* both published in 1938. The former had a second edition in 1943, and the latter ran to eight editions, the last in 1947, but no figures were given for the print runs.

If there is any doubt in William Herrod-Hempsall's worship of Thomas William Cowan and hence his determination to protect the legacy and reputation of the man, reference to the following which appeared at the end of the introduction to Herrod-Hempsall's guide should dispel any doubts:

IN MEMORIAM

Clarum et veneabile nomen.

Thomas William Cowan, F. L. S., F. G. S., F. R. M. S., F. E. S., D. Sc., Ph.D. : honoured and respected in every country where the honey-bee is cultivated the doyen of English bee-keeping, scientific, literary, and practical: to whose works, personal advice and assistance the author hereby reverently, affectionately, and gratefully acknowledges his indebtedness for the apicultural knowledge acquired or any success attained.

Unfortunately for historians, it also makes him an unreliable witness.

The 'slanging' over this squabble continued unabated a long time after both Cowan and Digges were dead, most contributors playing to their own audience. Cowan warranted no obituary notice in the *I. B. J.,* likewise Digges in the *B. B. J.*

To illustrate the level to which this battle stooped, Cowan claimed that, in 1904 one of Digges associates, Rev. Meehan, had been in court for poaching on a Sunday with gun and dog on the preserves of Lord Leitrim. James K. Watson informs that Digges was chaplain to the Clements family, owners of Lough Rynn Castle and Estate, formally owned by Lord Leitrim. Digges lived in Clooncahir Rectory, his church being Farnaught, and built by Lord Leitrim on that estate.

As a further example, consider an incident from 1925. Digges had written several times that the advice given in the governments' "collected leaflets" in respect of creating an increase was not suitable, each time the language becoming more intemperate. It was a much repeated method of increasing two colonies to three, resulting in one hive with queen, brood and young bees; another with foundation or supplied comb, queen and young bees; the third hive with brood and older flying bees where the new queen was to be bred. The idea was not new, Cheshire had it in *Practical Beekeeping* 1878, and there are elements of it in the first edition of Langstroth, although difficult to isolate in the narrative. Cowan also had it in his *Guide Book*. Digges dislike of the method was because the queen-rearing colony was initially stocked with older bees that had been drifted from one of the others. Interestingly, the method recommended for artificial swarming; creating two colonies from one; was the same as generally used today, resulting in the flying bees and queen in one hive on the original stand, with the brood and young bees, in the other. We now also drift bees from the queen rearing colony to the swarm, using one of many methods available, but this was not practised in the early 1900's.

Ironically, *The Irish Bee Guide* also carried the same method for creating three colonies from two! This double standard was pointed out in a letter to the *B. B. J.* from "A Loyal and Grateful Irishman, Cork", in late 1925. It had been known for the editors to 'plant' such letters on previous occasions, or ask someone to produce it for them, always refusing to give names when challenged; the writing style gives rise to suspicion in this instance. The letter was headed "Satan Rebuking Sin". Two weeks later the heading reappeared above a very much longer entry from W. Herrod-Hempsall. The opening paragraph contained the sentence:

```
It (the outrage from Digges) was merely mud slinging,
and I am delighted to see that his own countryman has
converted the missile into a boomerang; it will need a
good brush and much labour to remove the stain from the
rev. gentleman's vestments.
```

He follows this with much of the usual cynicism. Typically for W. Herrod-Hempsall, he concluded with:

```
We fear that with all this ingeneousness he is in the same
predicament as that other Irishman, who having dilated upon
the wonders of the sun and its circuit of the sky, when
asked if, as he stated, it rose each morning in the east
and set again in the west, how it got back to the east again
replied: 'Begorra, he just waits till it is quite dark and
```

> every one in bed, then steals back in the night time'.

A completely outrageous fabrication even for the time, and presumably something that did not apply to the "Loyal and Grateful Irishman" of Cork! Ireland had been independent for a few years at the time, but it was still difficult for some to accept; one can only guess at his master's feelings about it. W. Herrod-Hempsall followed it with more such material a few weeks later, allegedly by request. It was low-grade schoolboy humour, and should not have had a place in the *B. B. J.* An elderly local beekeeper remembers W. Herrod-Hempsall, when talking in Newcastle upon Tyne during the late 1940s, referring to "the Rev. gentleman in stained vestments" and using the term "Satan rebuking sin". The scars remained.

The Ministry leaflet continued recommending the method; the entries in the two guide books continued unchanged, that in Digges survived his death and Manley's complete takeover. It is unlikely that the *B. B. J.* articles were not seen by Digges, but he appears to have simply ignored them, refusing to recognise the contradiction that he was perpetuating, or it was possible that Manley was more than just advising by now - he is not credited in the 1921 edition, but was the sixth of 1928. As with most things in this dispute, logic did not apply, this was payback as Digges saw it, for the criticism by Cowan of the public money spent supporting Irish bee-keeping earlier.

One of the things at the heart of the problems was the desire of Cowan/Herrod-Hempsall/BBKA for legislation to deal with bee disease, and others, especially the bee farmers, very much opposed to it. Although Digges favoured legislation, the manner of his involvement and the illogical manner that it developed is detailed elsewhere. The matter of bee legislation had been an open sore for many years, and Hewitt's dispute with Cowan also contained aspects of it. Hewitt openly accused Cowan and his followers of wanting legislation to destroy colonies with foul brood, because they could not deal with it in their own apiaries. The cures that they sold were ineffective, and they wanted rid of his Punics, the last of these requiring a real leap of faith. In the *I. B. J.* of September 1921 there was a letter about the bee legislation stating that:

> ...the people keen to see it implemented were the 'Herrodians', and they look up to one who is a king among them, great and mighty and all powerful, whose word is regarded as law. They are beekeepers who cannot keep their own apiaries free from disease, but are most anxious to help other bee-keepers to do so.

It was the same thing that Hewitt had baited Cowan with several years past, but on this occasion he did not respond, having previously been badly beaten by Hogg on behalf of Hewitt, something that is also covered elsewhere. The Herrod-Hempsalls, however, did not hold back, claiming that the letter-reflected on their abilities as expert bee-keepers. But was it a libel? The letter writer apologised, donated to charity and paid costs. Digges did the same, but it could hardly be considered a genuine apology:

> ….apology for the insertion of the article in question, and to add that had we not overlooked the possibility of such an application as the article evidently lent itself to, we should certainly not have published it…...

But I bet they thought it was worth every penny! However, yet again there is only one source for this story.

There were many letters of complaint to the *B. B. J.* about the articles written on Cowan vrs Digges and those that preceded it on *The B. B. K. A. and it's critics*, a related item, feeling that the matter should have been left dormant. The Herrod-Hempsalls responded that 'their' side of the story – 'the truth', needed to be told and the dissenters were the opposition, but did not publish any of the letters. It was a very spirited and forthright defence, Cowan would have been proud!

As usual one is left with many unanswered questions. The information could have been published in 1904 when it would have had greater impact, and if Cowan felt he was being abused following the dispute, surely making the information public would have dampened that at least. I do not believe that it was because of his benevolent nature, as claimed by Herrod-Hempsall, otherwise why insist on it being published after both principals were dead? if indeed, he did. So one is left to speculate that by doing it in this manner he ensured that Digges could not respond, and his own supporters would hear the story only from their leader. However, the Herrod-Hempsalls only used it when they were under attack, some three years after the death of both Digges and his magazine.

Digges only apologised for copying thirteen diagrams, the text was not included, possibly because it was more difficult to legally prove, and of Herrod-Hempsall's own admission, some of the conditions were never, or only partially, executed. So why did Cowan not follow this up? He was obviously keen to see all the conditions in the agreement in the first place and it would have provided even more column inches for Herrod-Hempsall. Once again one has to question the accuracy of Herrod-Henpsall.

When studying this dispute, I was initially surprised at the ease with which Digges allegedly capitulated, but he was guilty as charged, and was left to negotiate the best deal possible and walk away knowing that he could fight another day. As well as the not insignificant compensation of £25, Cowan had the great satisfaction that Digges acknowledged that all his bee-keeping knowledge had been gained from Cowan's books. Cowan did not know at the time that it would only appear in a few thousand books of one title. I have suspicions that there was a second document, a 'gagging agreement', signed by both, that Herrod-Hempsall or Cowan himself have not admitted and made public, that might have explained some of the actions or inactions following the legal case.

The obvious hatred that each felt for the other, continued for as long as they lived, and an outright winner of the battle was never likely to emerge. But I believe Digges to have been the more intelligent, a better writer and shrewder operator. He was also a successful businessman on several fronts, whereas Cowan viewed himself as a philanthropist. His lack of business ability was exposed on many occasions, but his knowledge of bees and bee-keeping was far superior to that of Digges.

Considering that he operated, with only the aid of a secretary, from a rectory in the depths of the Irish countryside, Digges achieved much. As with his opponent he was well respected amongst his followers, and whilst he did not die a very wealthy man he once remarked that he "did very nicely". The two men appear to be totally different characters, but neither can be relied upon for the truth.

It is believed that R. O. B. Manley had much information from Digges on this matter, and he was of the opinion that Digges could have mounted a better defence. He claimed that Digges countered by accusing Cowan of doing the same re Cheshire's works. It is certainly true that Cowan added acknowledgements to Cheshire in the 1907 issue of the *Guide Book* that did not appear in the 1903 edition, even though the diagrams did, strengthening that particular argument. Annie Betts claimed that she had accessed this material, and wrote in *Bee World,* June 1937, under the heading "A Fallen Leader". For the period, Betts used quite intemperate language, and challenged the "Herrodians" to respond with evidence showing that what she had written was inaccurate. I have not been able to find such a response, presumably, she would have been included in "the opposition". Betts had a habit of making multiple carbon copies of letters and circulating them to her friends. I have several examples of these found stuck in books, a habit applauded by every historian, but deplored by the bibliophile in me. One of her letters (to Manley) relates to the quarrel retold here and was probably Manley's source for his statement above, and led to the article in *Bee World*.

Whatever one's feelings concerning an individual it is always sad to find, again attached to a book, the following, from Joseph Herrod-Hempsall about his brother William, dated 29th Aug. 1951:

> ......am having a bit of worry lately especially about Will.
> He is not getting on very well, can't digest his food very
> well and the doctor says that will probably get worse and
> he has also got colitis which causes him a bit of pain at
> times. He is at home, with his bed downstairs. He can
> manage to get out of bed sometimes but has fallen several
> times when doing so. His wife is feeling the strain so
> my wife has been over at nights to stay with her, as they
> could not get anyone to stay with them. .... I think he
> is gradually getting weaker. I am afraid it is only a
> question of time before he reaches the end of the road.

At the end of the book there is a further communication inserted from the same source, dated 11th September 1951, with the information that his brother died "Saturday morning last" - a very personal record of the passing of a dynasty. Unfortunately it did nothing to confirm or otherwise, the apocryphal story of the alleged occurrences the day following his death.

By way of a footnote to this quarrel, I offer the following taken from Cowan's *British Bee-Keepers Association Jubilee*:

> At this meeting the Hon. Secretary reported that in
> accordance with the power vested in him by the Committee,
> he had solicited and obtained the consent of the Baroness
> Burdett Coutts to be President of the Association , a fact
> that elicited a joyous hum from those who had so long been
> queenless ; and consequently profitless.

And from the *B. B. J.* editorial, May 1st 1878:

> The Hon. Sec., the Rev. H. R. Peel, reported that in
> accordance with the power vested in him by the Committee,
> he had solicited, and had obtained the consent of, the
> Baroness Burdett-Coutts to be President of the Association,
> a fact that elicited a joyous hum from those who had so long
> been queenless; and we may say, consequently profitless.

Cowan was <u>not</u> editor at the time. It is, of course not identical, but neither were Digges lifts; one might excuse Cowan on age, or perhaps by now he believed that he had always owned and edited the *B. B. J.* maybe he did actually write it, even though Peel was secretary of the B. B. K. A. and Abbott editor of the *B. B. J.* at the time, making the omission of Peel's name in Cowan's version more of a Freudian slip than a deliberate omission?

Unlike Hewitt, who had to create a magazine to be heard and to try to make Cowan's life difficult, Digges was already there and with a ready-made audience. When Digges published *The Irish Bee Guide* he did no more than all authors of textbooks before and after him.

Chapter X

The American Experiment

The above advertisement appeared in the B. B. J. of July 23rd, 1896 when Cowan was still smarting over the Punic bee fiasco and was being taunted by Hewitt. He was also witnessing increasing criticism of the central Association, and it had been certain for some time that the second attempt to obtain legislation to deal with bee disease would be a failure. On the personal front things did not appear much better. Assuming the reasons usually given for his non-attendance at meetings to be accurate, Mrs. Cowan suffered ongoing health problems; neither of his daughters had married; and only one of his sons, Percy, had managed gainful employment, and hence some degree of independence. Alexander the eldest at twenty-six, was now in California running a fruit farm set up by his father although at twenty-one he was allegedly an Electrical student. Career progression?

It was clear, that Cowan was relinquishing his lease on Penleaze at Fowey, the only place where he still kept bees, and it would appear that it was only three stocks at that. Three issues of the *B. B. J.* later, a similar advertisement appeared but with many of the items missing, presumably sold.

At the conclusion of the usual conversazione held during Dairy Show week, on October 22nd, a member announced that:

...the Chairman was understood to be starting shortly on a long journey—namely, to California—and that he would probably be spending some months on the continent of America. No doubt he would while there be on the look-out for everything likely to interest bee-keepers, and to promote the bee-keeping industry. He felt sure that the meeting would like to join him in wishing their chairman, and also Mrs. Cowan who, he understood, would accompany her husband—a prosperous voyage and a safe return to their many friends in this country. He asked the company to express their concurrence in that proposition.

The audience rose and loudly applauded.

The Chairman responded saying that he:

...was very much obliged by the kind response to the motion. He was never so happy as when amongst bee-keepers, and was intensely gratified to see that evening so many more present than usual. Mrs. Cowan and himself were going to spend the winter in California, where one of his intended tasks was to start an apiary for an Agricultural College in a fruit-growing county, some of the members of which had invited him to do so. He had already sent out books and other information on the subject. First of all he would make himself acquainted with the resources of the county, and if they were suitable he would be glad to start the College in so useful an industry, and give the necessary advice. He was quite sure that wherever he went bee-keepers would receive him as the representative of Great Britain. The last time he visited the States and Canada he received a hearty welcome, and was treated with the greatest cordiality, which he did not doubt would be repeated. Probably he would not be at the service of the College till the spring, but whatever he saw that was new and of interest would be treasured until he met them again.

I was unable to find any further reference to the College or it's proposed new apiary.

Cowan donated £10 to the Association funds and with Mrs. Cowan departed for America aboard the SS Etruria on November 14th 1896, travelling to their son's fruit farm in Loomis, Placer County, where they spent the greater part of the Winter.

In March 1897, the *American Bee Journal* reported that Cowan had been communicating with American bee-keepers seeking stocks of Italians for his son's farm. In the letter he stated that his son had never kept bees, so he believed pure bred bees would be better to handle than hybrids. They were required for pollination and adds "… I thought of Italians, as I know they are a favourite bee in America, although with us they do not do as well as the pure bred English bee". Apparently there was only one bee-keeper in Loomis and he destroyed his bees to take the honey.

In the same month Cowan read a paper, *Bees in Relation to Horticulture*, to the Fruit Growers Convention in Los Angeles, although he was less than pleased with the time he was allocated; not like the B. B. K. A. meetings, then.

June witnessed the publication of the fourteenth edition of the *Guide Book*, his address still being given as Belsize Gardens.

In Spring the Cowans were in Oakland, California, a West coast city on the San Francisco Bay, where fellow bee-keeper, W. A. Pryal showed them around. Cowan imparted the information that they intended spending next Winter in that part of the State and visit bee ranches in the southeast portion, before taking a trip east. According to C. P. Dadant, a bee appliance dealer, Mr. and Mrs. Cowan were spending most of the summer in Monterey "a quaint old sleepy town", but also spending a period in Pacific Grove, an adjoining settlement. Thomas had spoken at the University Farmer's Institute (University of California), on *Bees and Flowers*, also at Pomona College, Claremont, a recently established private facility, where Prof. Cook was now working.

Initially 1898, was to prove a relatively uneventful year for Cowan in particular and bee-keeping in general. W. Broughton Carr started the *British Bee Journal* year with a piece on his senior editor including, for the first time ever, a portrait. Cowan had given the junior editor a 'free hand' to do as he thought best with the Journal and its affairs whilst he was away, but as he was still in the employ of Cowan he was highly unlikely to jeopardise his situation; his boss must have provided the photograph.

In January 1898 Cowan attended California State B. K. A. convention, reading a paper on *Bee-keeping in England*. Unfortunately, he did not provide a copy of the lecture to the *American Bee Journal* and we have to rely on a summary by John H. Martin. It can be assumed that the report must have been reasonably accurate because the presenter did not write to correct it, which he would most certainly have done if there were any errors. He started by comparing the sizes of apiaries in California and England:

```
…in England there are but few large apiaries, and large
apiaries are not encouraged; the small apiary is the rule
in England, and many of the apiaries are kept in fruit-
growing districts for the purpose of fertilizing the fruit-
```

blossoms. Bee-keeping, according to improved methods, began
about 1860. The Langstroth hive was introduced at that time,
and was used by the most progressive bee-keepers, but the
real advance did not become general with bee-keepers until
1873. Up to this time many straw and other rude hives were
used, but now the Langstroth hive is gradually superseding
all others.

Did he really not mention his own hive and the standard frame? He followed
outlining the organisation of the craft in England, the *British Bee Journal* and
the B. B. K. A, saying of the latter:

...... is doing excellent educational work, several books upon
the different branches of apiculture having been publisht.

He outlined the examination of experts adding: - Experts for
the handling of foul brood are appointed and compensated
according to the work done. The expert does not visit keepers
who are well up in their business, but it is the careless
or ignorant bee-keeper whose bees are found diseased. In
many districts in England the bees have all died from this
disease. It was virulent because many bee-keepers had no
knowledge of the interior of a bee-hive ; straw hives, or
something equally inaccessible, were in use ; such bee-
keepers would defy the expert, for there was no law to
compel the destruction of diseased bees. The Association
had adopted a system of payments where foul-broody colonies
were to be destroyed, and tho the payments were small it
satisfied the owner of the bees, and enabled him to purchase
healthy colonies if his own were all destroyed.

Not my understanding of the situation.
He continued by talking about the honey market in England:

A system of labels has been adopted. These labels are issued
by the county association. Each member is alloted a number
which is stampt upon his label ; if he sells inferior honey
it is traced to him, and his name is stricken from the
Association. The Association employs an analyzer, and if
a person is caught adulterating honey or selling it under
an Association label he is imprisoned; they are not let off
so easily in England as they are in this country.

I have no record of such a case, but I suspect that he was making a point.

Figures are produced for the British honey yield and that imported:

...American honey was held in good repute until in 1879 ;
in that year Mr. Hoge, who represented Thurber & Co., of
New York, sold a large amount of adulterated honey, and
American honey has not regained the prestige then lost.

The sources of honey in England are white clover, sainfoin,
linn, (ling?), buckwheat, and the heather honey of Scotland.
Heather honey is darker even than buckwheat, but it sells
for a better price on account of its delicious flavor;

I refer you to his remarks in England re heather honey, following a
conversazione paper. The adjective used was definitely not delicious. Thomas
continued by talking about the value of the bee as a pollinator, and told the
story of Lord Studely's farm:

An orchardist in Gloucester planted 200 acres of fruit; the
orchard was a complete failure in fruit-bearing until a
Scotch bee-keeper put in 50 colonies of bees. When properly
fertilized by the bees the orchard began to bear. The
acreage was then extended to 500 acres, and the apiary was
increast to 200 colonies; the orchard now produces a large
amount of fruit, a large jam factory is operated on the
tract, and all of this prosperity is owing to the beneficial
intervention of the honey-bee.

The 'Scotch' beekeeper referred to was John Smith who Cowan met at Lord
Studely's orchard – the story is related elsewhere. There was no reported
discussion of his paper, but Cowan contributed to that on foul brood repeating
part of what he had given in his paper adding:

There were bees in churches and other inaccessible places
which were liable to spread the disease. The church bees
were being removed, and in this thorough way of treating it
the disease was much better under control than formerly.
Mr. Cowan believed in destroying the hive and the frames,
for a germ of the disease would live in a crevice of a hive
for several years, and if at any time it became exposed the
colony was sure to become inoculated.

The report had been 'Americanised', I doubt if Cowan used such terms as
'Orchardist'.

Early 1898 the Cowans left Pacific Grove to return to England, calling on
several American beekeepers on the way. By late February they were at their

son's residence in Loomis and shortly after visited the Roots in Medina. The main purpose for this visit was to confront Amos on an entry in *Gleanings* by C. C. Miller on cell size that disagreed with Cowan's entry in *The Honey Bee* and appeared to give Cheshire as the authority on the subject.

Students of the honeybee had been fascinated by the shape, size and construction of bee cells since the seventeenth century, but it was only when foundation began to be manufactured en masse that the measurements became significant. In *The Honey Bee* Cowan had written at length on the history of investigations into this matter, finishing by detailing his own, attempting to convey that at last a scientist had done it. He did his investigations whilst in Switzerland one summer, and informs that he measured the angles using a goniometer - two rulers joined at one end; not likely to result in great accuracy. He does not reveal how he made the linear measurements. Root published the letter that Cowan hand delivered, and added his apology for not having studied *The Honey Bee* fully enough, doing the necessary penance by disparaging Cheshire. It is the standard Cowan – I have written the definitive word on this, there is no more to debate, just quote me. Unfortunately, the passage referred to contained a measurement of 0.02 inches for the dimension across the worker bee cell, when it should have been 0.2; dismissed as a typing error. However, the foundation manufacturers had long since decided upon the cell size that they would use – Root was using 5 inches across 24 worker cells.

Mr. and Mrs. Cowan were taken for a trip to the Root's Basswood apiary, during which the visitors agreed with reservations to a photograph being taken and, even more reluctantly, to it being published. It is a very rare informal image of the Cowans together and, an even rarer one of Mrs. Cowan, who is seated at the back of the carriage. The main man is standing in the foreground, of course. We were informed that the cat ate honey!

Root wrote of the Cowans:

Notwithstanding Mr. Cowan is the most talented and best informed bee-keeper living to-day, I believe, having traveled over nearly all of the civilized world; notwithstanding he reads eleven different languages, and speaks perhaps half as many; notwithstanding honors of various kinds have been conferred upon him by different societies for the advancement of science; notwithstanding he is editor of the British Bee Journal, and president of the British Bee-Keepers' Association, he is one of the most modest men I ever met. Unassuming and quiet in his manner, one cannot fail to be imprest that he is in the presence of a master of our pursuit.

The Cowans at Roots

Without doubt he has the most extensive library relating to bees of any man in the world. He has been all his life gathering together rare and old volumes pertaining to bees; and it does not make any difference what language they are printed in, he reads them just the same. He pronounced some of the bee-books we have, dating back two and three centuries, as being exceedingly rare and valuable—such as, for instance, Butler's Feminine Monarchy, printed in 1609, or two years before the publication of the common version of the Bible; also Hill's treatise on bees, printed in London in 1608.

When Mr. Cowan was here in 1887 he had with him his big microscope with which he has made some of his exhaustive researches. This instrument is one of the finest in the world, and was made by Mr. Cowan himself.

I found Mr. Cowan to be well verst in all the modern and ancient practices of bee-keeping. It seemed to amuse him that so many ideas were being discovered in these latter days that have been fully described in former works. For instance, the modern starvation cure for foul brood isfully described in Delia Rocca's works, written over a century ago, and printed in 1790. He was also amused at the way we Yankees have of inventing things that his countrymen invented and afterward discarded.

Mrs. Cowan is as simple and unassuming in her manners as
her husband; and the fact gradually begins to dawn on one,
when he begins to know her better, that she is a woman of
more than usual Intellectual attainments, ……. She is a
very earnest and able advocate of the doctrine that the ten
tribes of Israel, after their carrying away to Assyria,
133 years before the sacking of Jerusalem (when Judah and
Benjamin were taken to Babylon) were not lost by absorption
into other nations. She claims that the Bible justifies us in
believing that those Israelites migrated to Central Europe
and became the Anglo-Saxon nation; that England is Ephraim
in prophecy, and the United States represents Manasseh;
that the religion and principles of these two nations will
spread all over the world and become dominant.[9]

Reading Root's *An Eyewitness Account of Early American Beekeeping* it is impossible to judge the success of the Cowan's stay with the Roots. In one chair the English gentleman, no interest in commerce, enough money without work to live comfortably, and indulge in his hobbies. In the other chair the American, very active in business and the community, as he says himself mainly interested in 'piling the dollars'. Root started his beehive business in 1869, at the time if the Cowan's visit he employed over one hundred hands. The conversation must have been priceless! The photograph sums it up – 'not only have I been photographed with a child but also a honey eating cat'.

The Cowans stayed in Chicago being entertained by George W. York the editor of the *American Bee Journal*. They were given a tour of part of the city and visited the home apiary of Mr. L. Kreutzinger containing100 colonies. York reported that he found Mr. and Mrs. Cowan charming conversationalists, having travelled extensively and being so highly cultural. He reported that Mr. Cowan had a library of over 1000 volumes and carried an index with him, adding that the *Guide Book* sells 5000 copies a year; just out by a factor of two.

During the senior editors sojourn in America, Broughton Carr had written that he did not think British bee-keepers had anything to learn from Americans or Germans, a sentiment often expressed by his employer. Dr. C. C. Miller, Marengo, Illinois, objected in very strong terms. Broughton Carr responded by

---

9 The belief that Anglo-Saxon peoples of Britain and Northern Europe are directly genetically connected to the ten lost tribes of Israel gained popularity with the rise of the British Empire but lost when the Empire became the Commonwealth of Nations. So Fanny's time coincided with the peak of this belief. One of the leading figures was Edward Hine who was related to George Rawlinson the Canon of Canterbury and Professor of Ancient History at the University of Oxford thus naturally opposed to the doctrine. Interestingly in the 1881 census, the Cowan household registered Margaret Rawlinson, 31 years old and with no occupation as visitor. The Rawlinson family appear to have been close friends of the Michells, but I was unable to make a connection to George.

telling him that Cowan was in America and he should seek a meeting because he was only the junior editor in charge. The meeting duly took place as the Cowans made their way across America. Of the visit Dr. Miller wrote:

> Along with a strong desire there was also just a shade of dread of the coming of two such prominent people; but they brought with them a large stock of common sense and loving kindness, and immediately we were at ease. Mr. Cowan impresses one as a man with a wonderful fund of information, but who has hardly discovered yet that he is any better off in that respect than the ordinary mortal. It so happened that while he was here I had sent me a letter and a newspaper clipping in a foreign language. I didn't even know what was the language—thought by the looks of the printed part that possibly it might be modern Greek. But Mr. Cowan read it off at sight. It was Russian. It made me feel I'd like to begin life over again and be a linguist.
>
> Mrs. Cowan is so simple and unassuming in her manners that just at first one is hardly ready to recognize in her a woman of unusual intellectual attainments; but gradually the fact appears, and, along with very clear perceptions of the truth, she is possest with an intense zeal that others shall see the truth as she has learned to see it. My wife, who had more opportunity than I to become acquainted with her, gave as her verdict, "Mrs. Cowan is a woman that lives to do good." They are delightful people and above all they are good.

Apparently, Dr. and Mrs Miller had been taught to sing 'God save the Queen' with gusto by the Cowans. There was no information on whether in response, the Cowans were induced to sing 'Dot Happy Bee Man' or 'Beekeepers Reunion Song', both compositions by their host.

Having informed all those with whom they stayed with that they were so impressed by their residence in California that they intended returning in the fall to spend winter, Mr. and Mrs. Cowan sailed from New York on 11th June.

Cowan was straight into action at the R. A. S. show in Birmingham on 20th June, as a Judge, and he attended the B. B. K. A. meeting on the 8th July. Shortly after this he was off to Switzerland to meet up with Bertrand. Mrs Cowan and her two daughters went to Scotland. He was also at the meeting on 9th September where it was announced that he intended to be at the Dairy show on the 17th October to "say goodbye to us for a time although leaving on the 22nd."

He never made that meeting. I have covered the events associated with the tragic loss of two of his children and his movements at that time in the Chapter on Genealogy.

Upon arrival in New York, Mr. and Mrs. Cowan with their sole remaining daughter, Edith, travelled immediately to Loomis and their eldest son, naturally cancelling the proposed meetings with bee-keepers on the way.

It was not long before Mr. and Mrs. Cowan, with Edith, moved on to Pacific Grove. During their previous American visit they had said that they were so impressed by their residence in Pacific Grove that they intended returning the next fall. I believe that they were referring to Pinehurst, which they had stayed at during their visit, and had the full intention that they with three of their children were to make that the family home. It was early 1899.

Pinehurst was built in 1895 by George Brandt, who was interested in fine horses and built handsome riding stables in the grounds. He spent $4,000 building the house, and lived in grand style for several years using an inheritance to finance his lifestyle, but eventually ran out of money. It was the most impressive residence in the settlement at the time, had six acres of grounds and was described as a 'large beautiful Queen Anne house'. The property was taken over by the Anglo Californian Bank Ltd. to the sum of $3000 and passed into the ownership of T. W. Cowan in either 1901 or 1902 for the same sum. There is firm evidence that the Cowans were in occupation of Pinehurst in early 1899, and must therefore have initially leased it from the Bank. The illustration shows what an imposing residence it was.

Pinehurst, Pacific Grove, during filming of 'A Summer Place'

Not surprisingly, 1899 was a quiet year for the Cowans. There was little activity on the bee-keeping front. Thomas defended criticism by Cook on his views of honey dew; wrote an article, *Foul Brood Germs – Spores or Bacilli* for *Gleanings,* repeated in the *American Bee Journal;* wrote to the *A. B. J.* with a photograph of his new residence; wrote a very vague review of the new *ABC of Bee Culture,* and responded to a query on Napthaline following mention of it in his foul brood article.

In 1879 Pacific Grove became the home of the Californian branch of the Chautauqua Literary and Scientific Circle, an organisation involved in lifelong learning, and operator of a book club. Arising from the annual assemblies held by this organisation, a small museum was built to house the Natural History collections that had been donated. Naturally there was the usual lack of funds and no organisation to progress the museum. On October 20th, 1899 a meeting of 'friends' was held at the Cowan's residence resulting in the formation of the Pacific Grove Museum Association. In July the following year the Chautauqua Literary and Scientific Circle donated the Museum building and all its contents to the new Association on the understanding that they would become incorporated. At this meeting, Cowan, 'a noble type of man of broad scientific training and social culture accepted the Presidency of the Association.' Presumably by chance, he had found a new focus, and possibly his true vocation. Being a new Association there was unlikely to be the same degree of politics involved as he had experienced in the B. B. K. A., he was already the big man from the big house, and President. Quarterly meetings were held at Pinehurst. In England; very few British bee-keepers were ever invited to any of his residences and even when the B. B. K. A. Council were struggling for places to meet, Belsize Gardens was never mentioned. The President did exactly what was expected of him, donating natural history specimens, books and probably money to the museum, even though there is no mention of the latter. He was a regular speaker at their gatherings: -

1900. At the opening of the Museum he and another gave the history of the Association.

1901. Lectured, no title available. Mrs Cowan also spoke, but again unfortunately, there is no topic given.

1902. 'Picturesque Europe' illustrated with 80 views by stereopticon. The Cowan's holiday snaps! Edith contributed specimens from Santa Cruz, and Alexander described the Colleges of Cambridge.

1903. 'Enemies of trees'.

But nothing on bees.

In 1900 Cowan became a little more active in beekeeping matters. In

February, he supplied a paper to the Californian State Convention of Beekeepers on *Some Difficulties in Connection with Foul Brood.* It was a very brief effort, and just a variant of his well publicised views on the subject, containing nothing new. It was read in his absence, and initiated much debate upon the desire by most of the fifty present, to investigate the potential for legislation in that American State. Following the publication of the paper in the *American Bee Journal* there were a few subscriber's questions on some of the content, duly answered by the author.

It was widely publicised that Cowan would lecture at the Chicago Convention of Bee-keepers, but again he just supplied a paper and his apologies for not being able to be present. The paper this time was *Chemistry of Honey and How to detect it's Adulteration.* In his covering letter to Dr. A. B. Mason, secretary of U. S. B .K. A., he wrote:

> I have endeavoured to write it in such a way that every
> one should clearly understand what I mean, and hope it may
> be of some use in clearing up some of the confusion that
> exists in a few minds about honey, glucose and sugar.

After the reading of the paper, A. I. Root who was chairing the meeting said "This is a most valuable paper, but it is a little too technical for most of us, I imagine" and promptly moved on, not exactly what the author expected! It was obviously not too technical for some of the *American Bee Journal* readers, and 1901 issues carried considerable criticism of some of the content of the paper; confidently answered as always. When re-printed in the British contemporary it received a much warmer reception, as would be expected.

It was becoming evident that American bee-keepers were not prepared to accept all that Cowan wrote just because he had written it. Since the Hewitt debacle the editors of all the American publications had approached Cowan's material with much more of an open mind, and were very ready to carry criticism, not countenanced in Newman's day at the *A. B. J.* As he had come to live amongst them he was considered to be as much 'fair game' as their homegrown authors; America was a very different culture than Cowan had been used to. *The Honey-bee,* published initially in 1890, but slow to gain sales in America despite the generally good reviews, had received occasional specific criticisms.

Professor, A. J. Cook, was a highly rated American academic, who in 1869 had been appointed Professor of Entomology and Zoology at Mitchigan Agricultural College, and in 1893 obtained the chair of Entomology at Pomona College, an educational institution founded only six years earlier. He was responsible for inviting Cowan to deliver *Bees and Flowers* during his previous stay in California. This is the connection that has led many writers to believe, with W. Herrod-Hempsall, that Cowan was awarded an Honorary Doctorate, but the present day institution were unable to confirm this. Cook wrote the *Manual of the Apiary,* seen by many as the equivalent in America to Cowan's *Guide Book* in Britain. In 1888, Cowan reviewed Cook's book in his *British Bee Journal,* pointing out in great detail where he disagreed with what Cook had written. The review of Cowan's *Honey Bee* by Cook appeared in the *American Bee Journal* during the latter half of 1900, and might be considered as 'payback'. He itemised, complete with page numbers, things that were factually incorrect, or where he disagreed, and in some cases questioned Cowan's basic understanding of that specific item. Unlike the reviewer of *The Honey-bee* for the *British Bee Journal,* Cook had actually read the book, and even though he summed up by praising it, and prefacing his remarks as quoted quoted earlier, he added the curious line - "However, a small horse is quickly curried". It was a damming review and resulted in others joining in, and not just on the book in question. It was 'open season' on Cowan's offerings.

At the start of a letter where he was defending criticisms of his paper on the chemistry of honey Cowan wrote:

> Although, as a rule, I do not notice criticisms, recognising the right of every one to his opinions, and do not care to be drawn into any controversy respecting such criticism.

I have not been able to detect when this 'road to Damascus moment' occurred. He then continued to repeat the material in the paper, ending with:

> If, and when, I have the time and inclination I should like to criticise Prof. Cook's criticisms of my book, " The Honey-Bee," but altho I have a personal regard for him, I do not find that he has adduced anything which would cause me to alter any of my views exprest in that book.

Normal service had resumed. As far as I can determine he never found the time or the inclination. However, comparison of the first edition of *The Honey-bee*; the one Cook criticised; and the second that appeared a few years after the review, indicates that some of the criticisms were taken notice of. No acknowledgement was forthcoming, of course.

The whole Cowan family visited British Columbia in the summer of 1902, but otherwise Thomas was inactive outside Pacific Grove. Early in 1903 Thomas wrote to G. W. York, editor of the *A. B. J.* with the news that they expected to start out from California "about Easter" on their way to Boston, and thence to Europe and possibly Africa, and be away for a year. Two weeks later this had changed to one and a half years of travel being looked forward to. Were the feet itching, or was there some specific reason for crossing the Atlantic again? The mention of Africa is an indication that Hewitt was still bothering him.

The chartered members of the Pacific Grove Museum Association held a farewell reception on both the afternoon and evening of the 14th April in honour of Mr. and Mrs. Cowan. The resolutions were presented:

> In view of your unfailing devotion to and untiring efforts in promoting the interests of the Pacific Grove Museum Association, and withal, tour large-hearted liberality in its support;
>
> We, the officers and members of the Pacific Grove Museum Association do hereby express our high appreciation of your valuable service, and thank you most heartily for your many courtesies and kind hospitality in frequently entertaining the association at your home, and for that devotion and service without which this association could not have continued to the present.
>
> We extend to you our best wishes and invoke upon you the Divine protection, care and blessing during your proposed absence, and we express our hope and prayer that in the Providence of God we may, in due time, have the coveted pleasure of welcoming you back again to your beautiful home among the pines by the sea.
>
> On behalf of the Pacific Grove Museum Association ......

There were recitals, readings, dainty refreshments of ice cream and wafers, and Alexander presented:

> A most interesting and instructive exhibition of stereopticon views of the beautiful and grand scenery along the line of the Canadian Pacific Railway. ......Many of the views were beautifully coloured.

It was all reported in great detail by the local newspaper, and does not give the impression that locals thought the Cowan's absence would be temporary

There is little doubt that Cowan had made a big impression on and contribution to the Association. This was, without question the type of environment that he was well suited to; a fledgling organisation of which he was immediately President, enabling him to utilise his organisational and writing talents, but not involving a great amount of work or controversy. It was reported at the farewell event that the Cowans "…. planned an eighteen month visit to their old home in England." A journalistic use of 'old home' I presume, because they had given up their houses in Cornwall and London and had no 'home' to return to.

They did the usual thing of calling at prominent beekeepers on the journey to Boston honouring the pledge they made when not visiting them on the way out due to the deaths of their children and pleasantly breaking a very long overland journey.

Mr. and Mrs. Cowan arrived in Liverpool aboard the SS Saxonia on May 14th 1903. Edith, Alexander and Frances, travelled separately; ironically, Tilbury being their port of arrival.

The question naturally arises - why did they decide to return to England for what appeared to be a predetermined period, at this moment in time? Why eighteen months? They appeared to be nicely settled in Pacific Grove, Cowan wrote many times that Fanny's health was very much better there, they had been accepted by the local community, and clearly integrated well. Unlike in Britain where they rarely entertained associates in their home, they had held all the quarterly meetings of the Museum Association at Pinehurst, and had opened the gardens there at weekends and on holidays for young men to walk with their girlfriends. There appears to have been no controversy or conflict with those they associated with, and both Cowans seemed to have been very much 'at home' and their travelling desires had been well catered for. However, it appears that Alexander and his wife had given up on the fruit-farming venture, removing their need to remain and they, with Edith, had returned permanently to Britain.

Cowan made no attempt to keep bees in America, and seemed little inclined to travel to some of the bee-keeper's assemblies to which he was invited. Although most of the American beekeepers that he met were a different kind of individual with very different priorities than those he led in Britain, they were still much in awe of him, but unlike in his home country they were not afraid to criticise.

Could it be thatCowan's friends in England had been privately urging him to return home to deal with certain issues? It is certainly true that 'his B. B. K. A.' had done very little in his absence other than 'tick over', and the County Associations were, yet again, questioning what the central body was doing to justify the money that they were contributing. Foul Brood had been a continuing topic in the *British Bee Journal,* and a certain George Saunders was beginning to make waves on that front. Then there were the Irish in general, and J. G. Digges in particular. Had Broughton Carr been alerted to the, soon to be published, *Irish Bee Guide?* All this without considering that belligerent Yorkshire upstart, John Hewitt, who was still tramping about in the long grass refusing to go away or be quiet. There is little doubt in my mind that it was the undercurrents in England that persuaded Cowan that he needed to return home to sort them out. He believed that it would take eighteen months, but it was close to two years before he felt able to return to America one more time.

I have covered the major occurrences in the next twenty-three month period in two separate chapters – Bee Disease Legislation (third attempt) and The Digges Affair. The following are the other salient incidents.

The European Intermission.

Early 1903, Cowan wrote a couple of editorials for the *British Bee Journal,* on the old chestnut - liquid fruit sugar, and had clearly been communicating with a company producing it. His opinions on this were coming under scrutiny on both sides of the Atlantic. He also submitted his views on Lambrotte's paper on Foul Brood; presumably no other member of the B. B. K. A. had sufficient knowledge. On 20th May Cowan was back in London and chairing an Association committee meeting where he was welcomed back and responded expressing his gratitude and thanks:

> ...for the continued confidence shown by the council in electing
> him year after year to the chairmanship notwithstanding his
> absence from their deliberations.

He was later to reveal that he had proposed resignation, because he was not in the Country and unable to carry out the duties of Chairman of the Committee, but it was refused. It was, of course, just a suggestion and seeking reassurance. He never actually resigned, always asking the council to accept his offer of resignation. However, how an individual can hold a position, the sole purpose of which is to chair meetings, when he is permanently away is difficult to comprehend.

Following their return the Cowan family appear not to have had a permanent home, Cowan giving 10, Buckingham Street, London, the address of the *British Bee Journal,* as his contact address. It is unlikely that they lived at that address although not impossible.

After a short period the family moved to Pine Grove, Boscombe, Bournemouth. This property was part of the Shelly estate centred on Boscombe Manor, at the time owned by Lord Abinger, who, prior to his inheriting the Scarlet family Barony in 1903 was Captain Robert Shelly Scarlet.

Cowan judged at Lincoln show, and was nominated to represent the Association on the Council of the new Lady Warwick Hostel at Studley Castle (a horticultural training establishment for ladies), where he attended a few times. He was very active at the first conversazione following his return, and responded to the call to talk on his bee-keeping exploits whilst in America. It was quite short, but he complained that the Americans had claimed shook and forced swarming to be their idea, responding indignantly, "we have been doing just this for many years". Part of the summer was spent in Europe. It was as if he had never been away.

The *B. B. J.* editorial of February 18th, carried the announcement that there was to be an exhibition of Wild Flowers and Sea Weeds of California in Morley Hall, February 23rd. Flowers had been gathered, dried and mounted by Miss E. C. Cowan. "Entrance fee for the Young Women's Christian Association, in which she takes an active interest".

At the B. B. K. A., A. G. M. for 1904, there were proposals to change certain rules in an attempt for greater democracy and give the County Associations larger representation. The Chairman expressed strong opposition to electing members to the committee for 3 years, with a third retiring each year. The motion was defeated of course, and a slight tinkering giving more representation to Counties eventually passed. Cowan was in favour of Counties being represented by someone closer to London if the distance prevented attendance; after all he only ever intended it to be a London club for gentlemen, and often threatened to leave if the rules were changed to make it more democratic, stating that it had to remain a philanthropic organisation for him to continue membership. However, he never explained why he considered democracy and philanthropy to be mutually exclusive.

Cowan was asked, and duly obliged to give evidence to the Department Committee appointed by Lord Onslow of the Board of Agriculture to report on the fruit industry. He was well suited to the task, as he was when the Board of Agriculture requested leaflets on *The Preparation of Honey for Market* and *Beekeeping for Beginners* to be prepared for issue by the Department.

When Rev. R. M. Lamb wrote to the *B. B. J.* complaining that he had been mis-quoted in the report of a conversazione, stating that he had said precisely the opposite, to illustrate that America had not really mellowed him in any way, the Senior Editor responded:

> Without knowing what our reverend correspondent had in mind when speaking, we assure him the words printed in our columns are reported verbatim. Moreover we can personally confirm their accuracy, having heard them.

It was never clear if the *B. B. J.* or B. B. K. A. hired someone to take verbatim the discussion at such meetings, the view expressed appeared to be whatever suited at the time.

At the March, 1905 meeting of the Council and A. G. M., the Chairman announced that he and Mrs. Cowan would be leaving Liverpool on board the SS Saxonia on 11th April bound for Boston; thence to their home in California, adding that:

> ...it was always a pleasure for him to attend this meeting, especially as members were ever ready to forgive any shortcomings......he had already taken a return ticket only available for one year.

Greeted by loud cheers – the ticket details presumably, rather than the newly found humility. On the return leg they were planning to visit John Hooker, who was in Philadelphia at the time; thence to Canada and British Columbia before sailing.

By the time the Cowans had arrived in Boston on 21st April the plans had been changed and they visited Hooker in Philadelphia, then to Cleveland, Chicago, Denver, and Salt Lake City on the way to their house in California. When he visited Dr. E. F. Phillips at Pennsylvania University, and E. L. Pratt at his apiary in Swathmore, he was carrying some coloured drawings of the anatomy of the honey-bee, which both Americans admired as reported in the *British Bee Journal*, but it is not made clear if these were his drawings or those published by the B. B. K. A.. In 1891, Cowan at the height of the Punic bees controversy, had written an editorial vilifying Pratt and his associate at that time, Henry Alley, headed "Humbug in the bee trade". Hence, Pratt appears to have been a strange choice for Hooker to arrange for Cowan to visit.

They visited the Roots as usual, but he must have been aware that Amos had written that the New York bacteriologists had stated that his foul brood was not the same as that in Europe, although that was precisely what Cowan had declared when he viewed it through his microscope in 1887. Despite their past skirmishes he also met with Cook whilst in California of which the latter wrote:

> Those of us who know Mr Cowan cannot but appreciate his great ability and equal courtesy. It is good to know a man who is so thoroughly and invariably a gentleman. Like all great men, Mr Cowan is an example of modesty. He never pushes himself to the front. His book *The Honey-bee* is one of the very best written by anyone,..... It is a model of accuracy so that no one need question any statements of facts taken from this work.

Oh for an accurate report of the meeting. Presumably he was now referring to the new, corrected, edition of the book. After it was printed in the *British*

Bee Journal Broughton Carr added: - "Comment on the above eulogy would be superfluous." The original Cook review of *The Honey-bee* had not appeared in the *British Bee Journal*. Of course not!

The Cowan's left their house in Pacific Grove for the last time on September 9th 1905, headed to Portland to visit the exposition, and then on to Canada and eventually travelled from Vancouver on the way to Boston, where they sailed aboard the SS Ivernia on the 21st November, expressing pleasure that they would be in England in time to spend Christmas with all the family.

Even though it was 1912 before Cowan sold Pinehurst for $600 more than he paid for it, the American adventure was over; a decision clearly made before they travelled back to England in 1903. Thomas and Fanny were now both past sixty-five, and their days of travelling, which they expressly enjoyed, was mostly behind them. Given the much stated health benefits for Fanny and their great enjoyment of the community in Pacific Grove, the question still remains – why did they return to England?

Chapter XI

Publications

Thomas William Cowan was a prolific writer on bees and beekeeping. He authored five books and a large number of leaflets and magazine articles. Additionally, for forty years he was the proprietor and editor of the only British magazines dedicated to beekeeping, but his editorial input diminished after he employed W. Broughton Carr in 1890.

His books were intended to inform and educate. Personally, I found most of his writing to be clear and concise, avoiding unnecessary complication or elaboration, an indication of his clarity of thought and command of the subject. However, in Cowan's case, this also led to him closing his mind to any alternative views, leading some to label his style of writing as 'arrogant'. Generally his books were well received, but it is difficult to accurately gauge this.

Cowan's first published efforts appear in *English Mechanic* and were associated with science and engineering and not beekeeping. I have written of these in the chapter on his industrial activities, but would reiterate here that they set the tone for his magazine writing - he had done it and described how, if others were not successful doing it, that was their fault, - they had not followed the instructions correctly; had used the wrong materials etc.

Major works.
1881. *British Bee-keeper's Guide Book to the Management of Bees in Moveable Comb Hives and the use of the Extractor.* Published in London by Houlston and Sons, Paternoster Square.

The first public notice that the book had been written appeared in the June 1881 issue of the *B. B. J.* In an advertisement headed "In the Press, and will be ready shortly", the title is given as above and it's author, followed by, "With numerous illustrations. Fcap, 8vo., price 1s 6d; or in cloth gilt, 2s 6d. To be had of all hive dealers and Booksellers." The July edition carried the same entry but headed "Will be ready, July 15th", and booksellers replaced by the publisher. By August the heading was **NOW READY,** "Postage 2d" had been added and a review appeared:

> This eminently practical and useful work, written solely in the interest of amateur beekeepers, will take a place in the front rank of British bee literature, and will retain it, too, deservedly. Mr. Cowan and his readers may be congratulated on the appearance of a volume which bears

```
on every page the direct impress of a master mind in bee
culture, uncontaminated by the egotism which too often
prompts to the publication of narrow views and 'one-eyed'
catalogues of goods for sale. Free from technicalities,
and 'phine phlights of phancy,'which pall upon the taste
and sicken the understanding, every line in the book may be
read with pleasure and profit, and re-read with increased
interest. We have only good words for a good book by
a thoroughly practical and liberal author; it is easily
written and well illustrated, and will be for all bee-
keepers a reliable guide and agreeable companion. - Ed. B.
B. J.
```

He liked it! One can only speculate on how different this review would have been, a few years later, after the great 'falling out', and Abbott threatening to write a book of his own.

For book collectors, the gatherings (quires, sections), were stapled together, and bound with the covers using glue. This is inferior to, and cheaper than sewing and with age and moisture leads to rust marks on the paper from the staples. If they can be found the books are often partially or totally disbound, and I have found some, where the collation of the index and advertisements has been haphazard

Some of my collection of The British *Bee-keepers Guide Book*

There being no frontispiece the title page was the first, and was numbered i. The preface was next - pages iii to vi, followed by two pages listing contents – thirty four chapters and index. There were 128 pages of text and seven pages of index. The end of the book contained advertisements from G. Worldridge of Horsham who made and sold the Cowan hive, extractors etc.; Abbott Brothers; Pettitts; Alfred Rushbridge; Isaac Hale; The Great Hampshire Bee Farm; Dr. Pine; William Raitt; Andrew Blake; John Walton; W. W. Young;

George Neighbour; S. J. Baldwin and Houlston and Sons catalogue of Natural History; a very impressive assembly.

Thirty-four chapters and 128 pages gives an indication of the brevity of some of the entries, unfortunately leading to complete paragraphs conveying little or no useful information. This reflects his drive to omit all but that which he considered necessary.

As normal the Preface starts with his justification for writing the book:

> ...the constant inquiries I receive respecting the practical management of bees in moveable-comb hives is ample evidence that British Bee-keepers demand a treatise on Bee-culture, which divested of all verbiage and superfluities, shall plainly point the way to success.

He then proceeds to rubbish other volumes – contain superfluous material; too expensive; out of date; contain rash claims, and nothing more than a glorified catalogue – this included some of those in the list of advertisers at the back of the book. The illustrations are by blocks specifically prepared for the work, twenty done by copper etching, the others loaned by Abbott, Cheshire, Hooker, and Neighbour that illustrated their inventions. The preface was dated April 1881. Two thousand copies were printed, all of which sold by the end of September.

This criticism of other books is from the Chairman of the B. B. K. A. Committee that decided in 1879 that it needed it's own publication specifically aimed at cottagers. This was written by F. R. Cheshire and John Hunter, both members of that committee at the time, and published in early 1880, individual copies costing 6d, 40/- for 100 and 35/- per 200. Hunter died before it was published. The preface appears to be in Cowan's hand, although not credited and includes:

> While believing that it (*Modern Bee-keeping*), will be found of great value, the Association does not desire that it should take the place of larger works, but rather that it should be considered an authorised introduction to them.

The first edition contained only 72 pages including all the preliminaries and index, and ten pages of advertisements. The engravings were supplied by Abbott, Cheshire and others and would become the centre of a major disagreement in a few years time. By the time of the ninth edition re-issue in 1903, eighty five thousand copies had been sold. The price would have made it attractive to cottagers compared to Cowan's offering.

British Bee Books a Bibliography 1500– 1976, informs that the second edition of Cowan's work was published in 1882, but the December 1ˢᵗ issue of the *B. B. J.* carries an advertisement for the book headed "**Second Edition. Fifth Thousand.**" It was identical to the first edition and in reality was just a re-issue.

As each edition sold, another was produced, some with only minor revisions, and others were just a re-issue. There were eventually a total of twenty-five 'editions', the last being in 1924, but there had been no revision after the nineteenth in 1907. R. O. B. Manley claimed that W. Herrod-Hempsall reissued the final edition several times after Cowan's death in order to limit the effect of *The Practical Bee Guide* by Rev. J. G. Digges (which Manley edited from the eighth edition on), and until he could produce his own book, *The bee-keepers guide to the management of bees in moveable comb hives* in 1938. But Manley did not like either Cowan or W. Herrod-Hempsall.

The three frontispiece used in the *Guide Book*

The first two editions did not have a frontispiece, after which there was an engraving of Cowan's gardener driving bees, this was changed in the fourteenth edition to a photograph of a gentleman, also driving bees, then in the fifteenth to a portrait of the author – by popular request, of course. From the fifteenth edition, the date was removed from the front cover, as was Cowan's name from the spine. The major revision of 1907 saw the front cover changed to match that of the second edition of *The Honey Bee*.

BRITISH
Bee-keeper's Guide Book
TO THE
Management of Bees in Moveable Comb Hives,
AND THE
USE OF THE EXTRACTOR.

ILLUSTRATED.

BY

THOS. WM. COWAN, F.G.S., F.R.M.S., ETC.

*Chairman Committee British Bee-keepers' Association;
Author of 'Bees and their Management,' 'Wintering Bees;' Member of the
Palæontographical Society, British Association for the Advancement
of Science, Quekett Microscopical Society, &c., &c.*

LONDON:
HOULSTON AND SONS,
PATERNOSTER SQUARE.

BRITISH
Bee-keeper's Guide Book
TO THE
Management of Bees in Movable-comb Hives,
AND THE USE OF MODERN BEE-APPLIANCES.
ILLUSTRATED.
BY
T. W. COWAN, F.L.S., F.G.S., F.R.M.S., &c.

Twenty-fourth Edition. Ninety-third Thousand.
LONDON
E. J. Larby, Ltd., 35, Paternoster Row, E.C.
and
"British Bee Journal" Office, 23, Bedford Street, Strand, W.C.2.
All Rights Reserved

Title pages of first and twenty-fourth editions of *Guide Book*

The title page changed with the passage of editions, and I show those from the first and twenty-fourth editions for comparison. The vignette of the bee only survived the first two editions, the hive being used thereafter. It is interesting to note that in the first edition, Cowan correctly referred to himself as Chairman Committee B. B. K. A. later this became incorrectly just Chairman B. B. K. A.

I believe *British Beekeepers Guide Book* to be Cowan at his best. It is practical, well-written, clear and very much to the point, if on occasion too brief to be of much value. Many chapters underwent no, or only very minor revision, and by 1926 it was well out of date. Unfortunately, he was incapable of avoiding the occasional embarrassment such as inclusion of the honey ripener and hive ventilation apparatus, neither of which managed to survive but a small number of editions. The book centres on his hive which was too heavy and expensive, and was soon overtaken by the W. B. C. hive which was very similar, but easier to handle and less expensive. The sales of the book were assisted by the never ending mention in the *B. B. J.* and *Record;* questions often being answered by simply referring the informant to the *Guide Book;* mentioning the book in correspondence was sure to get your contribution published, and it appeared to be an un-stated condition of an entry in 'Homes of the Honey Bee', was, for the author to say that he owed all his bee-keeping knowledge and success to the *Guide Book.* So the success of the book was never in doubt. It was translated into many other languages by his friends.

1884. *British bee-keepers practical notebook.* In line with the somewhat abnormal 'set-up' in beekeeping at the time, the first edition was published and available from John Huckle, at his home in Kings Langley, Hertfordshire. In 1884 Huckle was the paid secretary of the B. B. K. A. and the office manager of the *British Bee Journal,* for which I assume he was also paid. This publication was based upon Cowan's record book that was stolen at one of the London shows.

The first edition comprised just 56 pages and was bound in card covers. Following the preliminaries there were a series of tables for the beekeeper to enter data on Spring and Autumn observations; Winter observations; queen observations and rearing; number of brood combs; swarms; produce; feeding; autumn inspection; wintering; inventory; income and expenditure. It ended with four pages of very brief notes on seasonal management.

It appeared at the beginning of the bee-keeping year and was reviewed in the *B. B. J.* of 15th June, by his friend Peel, glowing, of course. In the introduction Cowan wrote that the book would:

```
...supply  a  long-felt  want,  frequently  expressed  to  the
author.   There  are  sufficient  Tables  for  one  year's  entries
in  a  large  apiary,  and  the  moderate  price  at  which  this
Note-book  is  issued  will  place  it  within  the  reach  of  every
bee-keeper.
```

It cost 1/- and required replacing each year; for comparison, his *Guide book* could be purchased for 1/6. It was not republished until 1904, presumably due to the author's absence from the country for much of the time, when chapters on making hives and bee-houses were added, partly contributed by others, taking the size to 73 pages. There was one further edition in 1908; it was a repeat of the second. I have found no information of the level of sales, but for the collector it is difficult to source, probably due to a combination of small sales and its intended purpose.

Two pages from the notebook are shown, an indication of the dedication Cowan wished to convey that he has, and that he expected in others. Running the risk that you dear reader, will now throw my book out of the window, here are a few pencilled entries from one of my copies of the notebook dated 1910: -

1 gross of bottles – 12s 6d

Carriage on bottles – 2s 2d

5 doz. Jar labels - 4s 0d

½ cwt. of s. sugar – 12s 0d

3 hives - £1 10s 0d

However, the entries also show that he was only getting 1s 0d per pound for his honey, the same as the price of the book.

14 I.–DAILY OBSERVATIONS FROM SPRING TO AUTUMN.

Month of   *Year*

Date.	Hour.	Height of Barometer. Thermom.	Wind.	General account of the Weather. Ingathering	What Flowers in bloom.

22   III.—OBSERVATIONS RESPECTING QUEENS.

Year

No. of Stock.	Description of Stock.	When Queen commenced to Lay. Month. Day.	Age of Queen	Race.	Size.	Colour.	Prolificness	General Remarks.

Sample pages from *Note Book*

1890. *The Honey Bee*. Published in London by Houlston and Sons. It comprises 11 pages of preliminaries, 192 pages of text, 7 pages describing the illustrations, strangely divorced from the engravings in the body of text, and finally a list of the works consulted, making, with the index, a total of 220 pages. It cost 2s 6d and was only published in cloth gilt.

In the preface, Cowan, as usual justifies the need for the book and that he has been "....repeatedly urged by friends to supply a long felt want....", -

familiar? However, unlike the *Guide Book,* he clearly states that it is a textbook and as such a compilation of other authors' work, although he claims that most of the illustrations had been drawn "...from the objects themselves, from our microscopical preparations, or from photo-micrographs taken by us specially for this purpose." He duly acknowledges the illustrations that are direct copies from other authors works. There is, of course no way of verifying how true this is; the head of a worker bee, will look the same no matter who draws it. However, a comparison of some of the diagrams with those in *Bees and Beekeeping, Vol. I,* by F. R. Cheshire, published four years earlier, bear more than a passing resemblance.

The different cloth covers of the first and second editions of *The Honey Bee*

R. A. Grimshaw, inventor of Apifuge, a lotion to rub on hands to discourage bees stinging them, reviewed the book in the *B. B. J.* Cowan had tried the preparation and reported that it was ineffective – got stung on the hands then! Others had the audacity to communicate that it worked for them, and the *B. B. J.* always carried an advertisement for it. Whether by coincidence or design, Grimshaw had written a series of articles in the *B. B. J.* on the development of the honeybee, appearing just prior to his review of Cowan's book. Grimshaw's articles had been heavily criticised, but he professed himself to be a 'non-evolutionist' on the matter. His review, which appeared in the last edition of the magazine for 1890, was typical of someone who had not read the book - not even the introductory remarks. It included:

> ...Far and away the finest book, the one containing the most true knowledge, yet published on the life-history of the

honey bee...Mr Cowan had apparently determined to disarm criticism beforehand by so checking his work over and over again, that error, even if it had crept in, could not possibly remain. ...a considerable amount of the information contained in the volume before me is, of course, the direct discovery and observation of the author himself and is both new and true... but with characteristic modesty the writer refrains from taking credit to himself.

One of my copies contained an errata slip, and in the preface to the second edition he openly stated errors in the first had been corrected in this the second.

The *American Bee Journal* and *Gleanings* both gave it favourable reviews, but some of the follow-up correspondence was not so complimentary, including a simple telling one liner "..tries very carefully to conceal the fact that he has never made any investigation himself".

Although it was a self-declared aim to produce updated editions in the same manner as the *Guide Book,* only one further edition was issued - in 1904. In the preface to the second edition, Cowan states that it has been revised and corrected. Forensic inspection reveals that he has rephrased small pieces of it, added the occasional paragraph, and included one additional illustration - Figure 11 in Edition two, which is simply an enlargement of part of Figure 51 in Edition one. Errors have been corrected but most of these were on the errata slip in edition one; some were those pointed out by Cook.

Comparison with Cheshire's book issued just four years earlier is inevitable because of the similar layout of the topic and the content. Although Cowan credited many, Cheshire is not included, and yet it is clear that he used his book; some of the text is very similar, and these were the passages that he re-wrote in the second edition. Annie Betts believed this was because of the controversy he was about to embark upon with Digges. Personally, I believe Cheshire's volume to be far superior to that of Cowan; the text is clearer, there are fewer lengthy pieces copied from other authors, the illustrations are better and their descriptions adjacent. Copies of Cowan's book are often found with the owners having written the descriptions on the diagrams themselves. Cheshire acknowledged his sources at the bottom of the relevant page.

There are no figures available for the sale of the book, but as would be expected for a scientific work, sales were slow. The second edition, to a large extent a reprint, appeared fourteen years after the first. As with the *Guide Book,* his friends translated it into German, French and Russian.

1908. *Wax Craft, All About Beeswax,* published by Samson Low, Marston and Co. Ltd. and The British Bee Journal. It was issued in paper covers at 2s 0d and cloth gilt at 3s 0d, the latter very difficult to source. Upcott Gill printed it on very poor quality paper, which is subject to foxing. The seventeen plates are on Art paper; there are 133 pages of text, 110 recipes, and an index. There was only the single edition.

Waxcraft was issued mid June 1908, accompanied by very large advertisements in the *B. B. J.* and *Record,* headed "Just out – new and important work". The first issue of the magazine for July had a 'Flyer' sent out with it, containing most of the title page, an extract from the Preface, (although in the book it was labelled 'Proem'), and the contents pages. The review appeared in D. M. McDonald's regular column 'Among the Bees'; two full columns of praise. The American bee press reviews of *Wax Craft* were reported in October, all complimentary again, "...cloth bound copies are the most artistic of any gotten up in all beedom".

In 1889 Cowan had edited and translated into English a paper by M. J. Dennler, - *Beeswax: it's economical uses and conversion into money,* so the basis of this book was there for him. He extracted every possible gram from the topic – very necessary in order to produce a full sized book, and, to my knowledge, it has never been superseded. Being a very specialist subject, it was never likely to find a large market amongst beekeepers, but I consider it to be an excellent volume, well worth reading, even today.

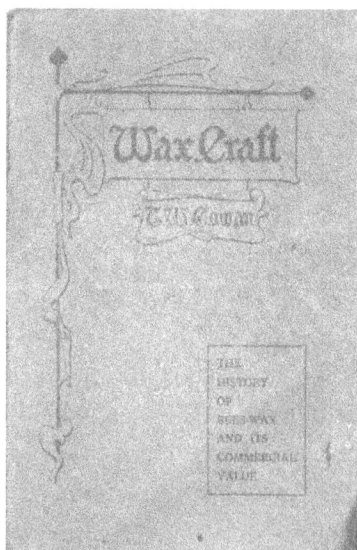

Beeswax, card covered edition and title page

1910. *The Queen Bee*. Published by his two magazines. 16 pages plus a dissectible model; 12" x 9", illustrated card covers. *British Bee Books* dates this as 1910, but all the copies that I have seen contain W. Herrod's book, *Producing, preparing, exhibiting and judging bee produce,* amongst the 'Books for Bee-keepers', in it, although with a slightly different title than it was eventually published under and lacking a price. This book did not appear until 1912. *The Queen Bee* is extracted from his book *The Honey Bee* but in the model, the text describing the various parts of the queen-bee is positioned alongside the relevant illustration. The 'dissectible model' is pasted inside the back page and is rarely found complete with all its loose parts. It comprises two cut out flaps that can be lifted up to reveal the various organs. Sometimes found with the loose parts glued in.

It is difficult to understand the reasoning behind the issue of this publication or its intended market. Seven years earlier, Gresham, the schoolbook publishers had issued something very similar, as part of a series including ants and spiders.

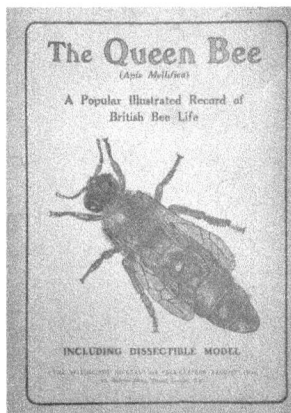

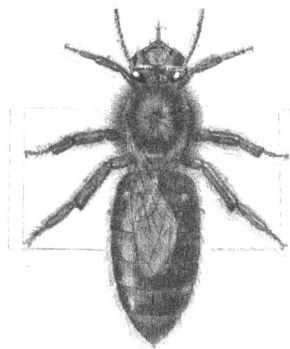

The queen bee, showing cover and fold out model

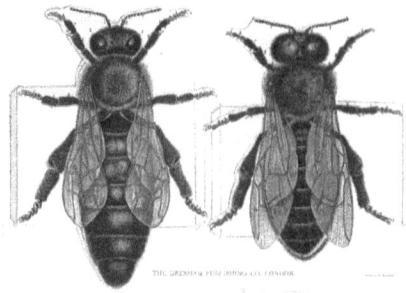

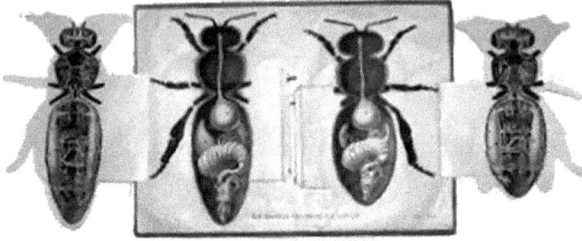

Bee model by publishers Gresham

I was unable to find a review or advertisement in any bee publication for Cowan's publication, which leads to the conclusion that it was intended for schools, rather than his usual readership. However, it is not listed in Natalie Hodgson's bibliography of children's bee books, and Col. Walker did not purchase it. It was subtitled '*A Popular Illustrated Record of British Bee Life*' but bee anatomy is never 'popular' amongst bee-keepers and it's difficulty in sourcing by collectors indicates it was not 'popular' on any level. Priced at 4/6, with postage 3 ½ d might explain the apparently, poor sales.

1928. *British Bee-keepers' Association Jubilee.* British Bee Journal, London. This had appeared in the *British Bee Journal* as a series of articles during the eighteen months prior to Cowan's death. What could have been an excellent factual account of the first fifty years of the B. B. K. A. by someone who was there for most of the time and held copies of the records, is a disjointed account by a very old man of those items he considered important. The parts where he did not feature strongly are generally omitted, and those that he was not proud of glossed over, modified or also omitted. Much space was given to his meetings with royalty, those in important positions, and visiting bee-keeping dignitaries.

There is very little evidence of the editing that allegedly took place, and the result is a disappointing opportunity missed. When I commenced my research for this work, I thought that this book was the obvious place to start; how wrong can one be!

Minor works.

Pre 1881. *Bees and their Management.* He claims authorship of this publication in the *Guide Book,* but this author has been unable to locate it.

1880. *Wintering Bees.* "The most complete work on the subject of Wintering published". B. B. K. A. publication, price 3d. As with most Association publications this was a reprint of the paper read by its author at a conversazione, and in this case followed his earlier contribution to the *B. B. J.* on bee management to obtain large surpluses, and one on the same topic in the *A. B. J.* reprinted in its British contemporary. It also formed the nucleus of the corresponding chapter in the *Guide Book.* The second edition appeared in October 1881 and the third in January 1886. The most popular of such publications.

1883. *Management of Straw Skeps.* "Designed to teach the Cottager how to obtain the best results at the least possible cost. By attention to its teachings, cottagers will be enabled to make bees a more profitable source of income than hitherto." B. B. K. A. publication, price 1d, cheaper in quantity.

This was pamphlet was the result of the following motion being passed at a Council meeting:

```
...that it is desirable for the B. B. K. A. to publish penny
pamphlets on the management of bees for the special use
of Cottagers, including one on the profitable use of straw
skeps.
```

It was the usual confused state, but the result was that *Modern Beekeeping* continued in the same format, with the illustration of the skep removed, and Cowan wrote this penny pamphlet on skeps. Not my understanding of the motion passed.

1887. *Doubling and Storifying for extracted and comb honey and the prevention of swarming.* 15 pages. 3½ d including postage. The advertisement for this pamphlet appeared at the very end of 1886, and was headed 'Guide Book Pamphlet No. 1' and was available from Huckle at Kings Langley. The *Guide Book* had a chapter on this topic of only three pages containing very little information but with a footnote in those editions after 1887 stating "For more information the reader is referred to my pamphlet ....".

1887. *How to Make an Extractor and Bellows Smoker.* Price " Post free, sixpence halfpenny".

A short article in the editorial columns of the *B. B. J.* preceded both parts of this publication. At the start of that for the extractor Cowan wrote ".....we do not think that it will generally pay an amateur to make one for himself...". On February 10th the, by now familiar build-up, "Shortly will be published No. II of Guide book pamphlets – *How to Make an Extractor and How to Make a*

Bellows Smoker. On March 31st it was 'Now ready' with the slightly modified title as above.

Issuing them as 'Guide Book pamphlets' I and II indicates that the intention was to publish many more, but presumably, the sales were not large enough to justify the effort. There is no evidence that there was any more than the single edition of either of them. However, they were both published complete in the *B. B. J.* during 1913.

1889. Bees*wax; its Economical uses and Conversion into Money.* A *B. B. J.*

Publication, price 3d. It was an article by J. Dennler that Cowan translated and edited. It appeared late July.

1895 *Photograph of Foul Brood. A B. B. J.* publication. Price, including postage 1/-. This was part of Cowan's response to the point made by the Minister of Agriculture, that foul brood could not be made a notifiable disease if bee-keepers could not recognise it in their hives. This photograph was reproduced in the *Guide Book* and is shown in the chapter on foul brood legislation. In later editions it was replaced by that shown below, but it is questionable if either was of much assistance.

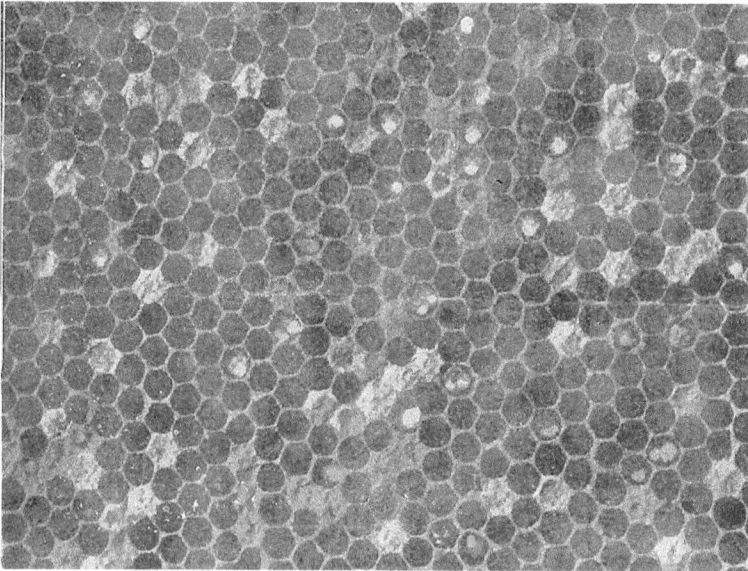

Foul brood photograph from later editions of *Guide Book*

1895. *Foul Brood and its Treatment.* This was an article Cowan wrote for the quarterly Journal of the Royal Agriculture Society and was reprinted in the *B. B. J.* Although some writers claim that this was published as an independent leaflet, a view supported by the *Guide Book,* - I have not been able to confirm this. However, as a result of the third attempt to obtain legislation to deal with bee diseases, the Board of Agriculture asked the Council of the B. B. K. A. to prepare a leaflet for them to distribute to beekeepers. This leaflet was written by Cowan and appeared towards the end of July 1896 as Board of Agriculture leaflet No. 32, with the title *Foul Brood or Bee Pest*. It was available from the Board free of charge and reprinted in the *B. B. J.* It was reissued in 1904, and revised for the first time in May 1907. The revision is substantially different than the initial Cowan version and includes replacements of his poor photograph, ironically, with two loaned for the purpose by the Department of Agriculture and Technical Instruction for Ireland.

1905. *Preparation of Honey for Market.* Issued by the Board of Agriculture and Fisheries as leaflet No. 141 in May. Although there is no author credited, this is almost certainly the work of Cowan, and he claimed such. Revised for the first time just two years later.

1905. *Advice to Beginners in Beekeeping.* As above but this was leaflet No.128.

1909. *Beneficial Results from the Fertilisation of Fruit Blossoms by Bees.* Published by the *B. B. J.* The result of several papers read on the subject in both Britain and America. Price 3d

Lectures at B. B. K. A. Quarterly Conversaziones.

1879. Wintering Bees.

1886. The Development of Bee-keeping as an Industry.

1887. Royal Jelly, Food of the Queen Larva.

1887. My Journey in America.

1889. The Choice of Hive.

1890. The Food of Larval Bees.

1890. Excursions in Savoie and Haute Servoie.

1891. Technical Instruction in Bee-keeping.

1892. Description of Hives, Bees and Photographs of Apiaries Brought from Africa.

1894. Demonstrated the Detection of Adulterated Honey by dialysis and the use of the polariscope.

1895. Showed and Described the First Australian Bees to be imported into Britain.

1903. Apicultural Experiences in California.

1905. Claustral Hive of Abbe Gouttefrangeas.

1906. Standard and other Frames.

1907. Brood Diseases of Bees.

1908. The Development of Bee-keeping as an Industry.

1908. The Influence of more than one Queen in a Hive – as stand-in for a French Bee-keeper that did not show.

Both 1908 presentations were at the Franco-British Congress of Bee-keepers

1909. Some Beneficial Results of the Fertilisation of Fruit Blossoms by Bees.

1909. Some Recent Investigations in Connection with Diseases of Bees.

1913. Bee-keeping in Other Countries Pt. I.

1914. Bee-keeping in Other Countries Pt. II – this followed his cancellation at an earlier date.

Neither of the previous two lists do justice to the volume or breadth of his writings in the *British Bee Journal, American Bee Journal, Journal of Horticulture, The Times,* and *The Field, etc.* Most of his written efforts were based upon his own personal experiences or reporting upon his study of other author's works; occasionally the boundary between the two was unclear. Unfortunately, like many such individuals he easily became involved in crusades, sometimes illogically; controversy was rarely far from his pen, and some of his written material suffered as a consequence. But when writing on basic bee-keeping operations etc. in my opinion he had few contemporary equals, especially when it did not contain the arrogance that unfortunately characterised him.

Chapter XII

Sun Setting

Thomas and Fanny Cowan returned to their temporary accommodation in Boscombe, November 1905, where Edith, Alexander and his wife had remained for the previous eight months.

Thomas reconnected with the B. B. K. A. at the February 1906 Council meeting, and attended the March A. G. M., chairing the quarterly meeting that followed. In his usual lengthy contribution to the discussion on a paper *The Italian Bee and Breeding Bees by Selection......"* he said he had thirty years experience of importing foreign bees, and the problem bee-keepers had with the Italians was that they used every one they bought, but in his experience only one in ten "was good for anything". He gave no indication of how he had arrived at that statistic or how they could be selected. Did he really think beekeepers would contemplate not using something they had paid for? He was also confident that foul brood was brought into his apiary by bees from Italy, but he had eradicated it because he had "a great many years experience in the treatment of foul brood". This tells us that he did not clear it at one attempt, but we now have the knowledge to know that he did not conquer it at all with his methods. His attack on Italians ends with "......the disease (is) very prevalent in Italy, although not so virulent in form as its English prototype", whatever that meant.

He continued by stating, yet again, that when breeding he chooses the "....very best queens and drones", but again fails to inform us as to how he does this. He considers the first cross of the Italian with the English to be better than the pure Italian. Another meaningless statement, because he does not define the characteristics which makes it 'better', or how he made the assessment. A few years earlier he had tried to source some pure Italians in America because he thought them 'better' for his beginner son. He now believed the best-imported bee to be the Carniolan, and declared the golden (yellow) bee, to be just a fancy bee, although he admitted to no experience with it.

The Chairman skipped the next item, moving swiftly onto discussing size of lace to be allowed when exhibiting sections, then to tasting heather honey samples, declaring that one was "....tasting more or less like heather honey"; very likely!

At the previous conversazione in October 1905, there had been discussion about the strength of the top bar of the British Standard frame, but with no definite outcome, although the same meeting felt confident to fix the recommended dimensions for a standard section. The reason for despatching the early items rapidly in the 1906 meeting was the desire of the Chairman to have time to return to the debate on the Standard frame. He commenced by speaking at length on the history of the decision, and gave the main reason for the choice of the size (14" x 7 15/16"), was that it would fit into the existing Woodbury hives, which were 14 ½", but he had given the size of the Woodbury frame earlier as 13 ¾". The decision had been unanimous. He had tried the Langstroth size, found them too large, and was trying German hives at the time the decision was made. He had come to the conclusion that there was:

> ...nothing to better the size finally adopted, which was <u>about</u> (!) the same dimensions as the present standard frame.

Having made the case for the status quo he said:

> ...... there was no desire to stifle discussion or invention; as they grew old in years possibly they were apt to look upon new inventions as not conducive to the best beekeeping.

However, when the Hoffman type frame was shown he said he had considerable experience of them "...and did not wish to have anything more to do with them" because of propolisation and crushing bees, which "might be acceptable in America but not here". He had always opposed frames with inbuilt methods of spacing, especially those introduced by C. N. Abbott, and those being shown were American; no chance!

Cowan appeared to concede the point that the saw cut in the top bar was part of the problem, but said that at the time that he introduced it German foundation was much in use. This was his first mention of German foundation. From the report it was clear that most of those present believed that the top bar needed strengthening, and an alternative to the saw cut was required to fix the foundation. The Chairman did not call for a resolution and vote, proposing that the frames shown be tested.

For the first time I detected some signs of disinterest and tiredness in Cowan. In many respects his performance was very un-Cowan like. The tone of most of his contributions was conciliatory, containing vague, sometimes meaningless statements and very occasionally, dare one say it, humility. Could it be that his enthusiasm was diminishing following his unexplained final return from California, and his absence from keeping bees for the last nine years? Might it simply be age and the diminished energy levels and subsequent mellowing that for most, accompany it? Could it be the dawn of a new non-confrontational period, the great leader realising that his Association's importance, in many areas, was diminishing, as the State became ever more involved and the County

Associations prospered? His Queen was dead as was many of his disciples; this Victorian Gentleman had outlived his era.

However, worse was to come. Late 1906 Baroness Burdett-Coutts died. Cowan was ill so did not attend the funeral in early January. The reason that he had previously given for remaining Chairman had gone. Shortly, the Association decided that its future Presidents would be the Worshipful Master of the Wax Chandlers Guild, an appointment that had a duration of two years.

Early 1907, the Cowan family, minus Percy, moved into Upcott House, Upcott, Nr. Bishops Hull, Taunton. He said he chose Somerset to live in because of its beauty and mild climate; he was talking to Somerset bee-keepers at the time. Somerset B. K. A. was delighted, of course, and made him a vice-president. Cowan did not travel to London for the final conversazione of 1907, but wrote enclosing a new steam heated uncapping knife, saying that he had been unable to try "the contrivance" because there was no honey that year; it had been a complete failure. The implication is that once again, he was keeping bees. He had also sent a sample of Mexican honey for comment and sampling. At the conclusion of the meeting it was decided that a note should be sent thanking him, accompanied by "general cheers". The adulation never waned!

Upcott House was originally a seventeenth century farmhouse, and situated across a lane from Upcott Manor House. It is reported that Cowan planted lime trees there for his bees.

Upcott House, present day

I have carefully avoided including much on the scientific developments associated with bee diseases, believing that it constitutes a book in its own right. However, by the turn of the century, 'proper' scientists were beginning to carry out detailed studies, producing results that were not totally compatible with those of Frank Cheshire in the 1880's. To most, this would be acceptable given the development of the various scientific techniques and the large effort being put into 'germ' theories. In the March, 1907 meeting, Cowan presented a paper, *Brood Diseases of Bees,* which was simply an attack on the work of Dr. Phillips in America and Dr. White in Britain. He reiterated his belief that there was two forms of foul brood – mild and virulent; but a further disease known as 'Black brood' had been found in America. He revealed that he first saw 'Black brood' eight or nine years ago when in America, and knew it was different than foul brood because it did not exhibit 'ropiness'. Phillips had stated that foul brood in America was not the same as foul brood in Europe, naming them 'American foul brood' and 'European foul brood'. Cowan was not pleased to put it mildly, he had not been consulted, he was being made to look silly, and the Americans were 'getting above themselves'. He claimed White's research was flawed.

This was another topic where Cowan considered himself the world number one authority. He had always been very positive in his diagnosis of diseased comb sent to him, but professional scientists were now questioning his judgement in both America and Britain. Staying with the Roots on his first visit to America, when he travelled with his wife and microscope, Cowan was presented with a diseased comb, and following inspection using his instrument, declared that it was foul brood, the same as in Europe. Root now knew differently.

Naturally, given his age and distance from London, the Chairman's attendance at Council meetings, began to decline, but he always attended the A. G. M., unless ill. He became more active in his local County B. K. A. He attended some of the larger shows to judge, always appearing if Royalty was expected, visited important exhibitions and usually represented the Association when giving evidence to government committees. His much deserved retirement as the Grand Old Man of British Bee-keeping continued to be punctuated with difficulties and controversies, which he seemed incapable of ignoring.

There are few that would not accept the concept of the B. B. K. A. establishing a system of examining candidates for various levels of competency in the craft that it represents. It is one of the enduring legacies of Cowan's tenure. However, by 1907, bee-keepers were beginning to question all aspects of the system that existed for examining and awarding three different levels of 'expert'. It had remained, virtually unchanged since its' inception, and very little was avoiding criticism. The answer was as usual, a sub-committee of the Association comprised of the usual suspects to deliberate. When Cowan presented to a departmental committee of the Board of Agriculture looking at the return on the money being spent on education and training in Agriculture, he wasted no time in criticising the appointment of some of those operating the scheme, because they were not B. B. K. A. experts. It is a view still held today by many 'hobby bee-keepers', who feel the government chief bee-man should also be a practical bee-keeper and a member of his local Association.

Whilst the criticism of the examination system had been temporarily silenced it was not long before the B. B. K. A. once again came under attack. At the coversazione following the A. G. M. in March 1908, E. Garcke, a member of the Council, read a paper, *Suggestions for Improving the Well-being of Beekeeping*. He had carried out a major investigation of the state of bee-keeping in Britain, quoting a large number of statistics that painted a very depressing picture of the state and organisation of the craft at that time, and he expressed great pessimism about the future. He was careful not to directly criticise the central Association, of which he was a member, but it was clear that its mission had not been accomplished. He proposed a conference of County Association representatives to seek solutions.

When Garcke finished speaking Cowan used his well-practised tactic of attempting to divert the meeting with another matter but those present were not to be silenced, saying that the status quo was not an option and the proposed conference was the way to proceed. Not for the first time, the Chairman's contribution was as long as the paper. In his customary manner he outlined the history of the subject as he saw it, justifying every step taken, 'the B. B. K. A. had done a splendid job, and the County Associations were reaping the rewards, benefiting from the government grants etc. etc'. It contains many quite extraordinary claims and statements, but he did admit to one mistake – fixing the B. B. K. A. subscription at 5/, because if it had been 10/- or £1 there would have been an increase in membership! There can be few better examples of the parallel Universe that Cowan inhabited. In his paper Garcke stated that just under 25% of the B. B. K. A.'s members paid £1 or more, but of course they received four votes by doing so. Even so, the Association still only had annual revenue of £125 (£140 given elsewhere). The Chairman stated that he was in favour of the conference if it resulted in strengthening the financial position of the B. B. K. A. It must have been clear to all present that the way forward would not be easy.

Most of us, have at some stage, been members of a club that was in desperate need of modernising, but in deference to the leading older members who did not want nor see the need for change, nothing was done. Opposition to change appears to be our primary function as we age. The situation at the B. B. K. A. was exactly this, and had been for some considerable time. The fear of upsetting the Chairman was uppermost in the minds of most members of the Council, but it was coupled with an even greater fear of not being able to manage without him, or not being elected to the Council. It is easy, with hindsight, to say that when he departed to America, they should have appointed a new Chairman, and suggested to Cowan that they would like him to act in an advisory capacity. Whilst he was not there someone else chaired the meetings; but it is also true that very little was done in his absence.

Cowan and Broughton Carr judging at Dairy Show 1908

Cowan was busy in a number of ways, *Waxcraft* had recently been published, he had issued the third edition of the *British Bee-keepers' Practical Note Book*, and redrawn many of the illustrations for the nineteenth edition of the *British Bee-keeper's Guide Book*, because Digges had purloined the previous ones. He had judged at several shows, spoken at the Congress of bee-keepers at the Franco British Exhibition in June on the *Development of Bee-keeping as an Industry*, and attended the Board of Agriculture again. He was active in Somerset B. K. A. and a member of Taunton Coversazione and Field Club as well as contributing a few articles to his magazines, it was easy for him to pretend that all was well in the bee-keeping world.

Very early in 1909 W. Broughton Carr died aged seventy-three. He had been the eyes, ears and mouthpiece of his employer whenever Cowan was away. He had been very loyal; his had been an excellent appointment. Given Broughton Carr's age, it is strange that the magazine proprietor had made no provision for a successor, and had to 'fill-in' himself. I have not encountered information on the help, that without doubt, he would have called upon to keep the magazines going, or whether he moved back to London on a full time basis, but he was ever present at the B. B. K. A. during 1909, which he probably thought no bad thing in view of the unrest.

Yet again, in response to the financial crisis in which it was in, the B. B. K. A. launched an appeal for funds, which brought forth correspondence in the *Journal* suggesting that a better approach would be reorganisation, which he surprisingly published.

On the 5th August 1909, the B. B. K. A. Council held a special meeting to consider Garcke's scheme. It was nearly eighteen months since his first paper and Garcke had now updated his figure for the revenue of the B. B. K. A. to £290 per annum, £150 being from 'other sources', but some of his mathematics leave a bit to be desired. He outlines the possible options as he sees them and concludes that the best arrangement would be for the B. B. K. A. to become a federation of the County Associations. The existing members of the central Association would pay their subscriptions to their respective County Associations, who would pass money back to the central body on the basis of a fixed sum plus a per capita amount. The B. B. K. A., in its existing form, would be disbanded and replaced by a body elected by all the members. The *British Bee Journal* informs that a "long and detailed discussion followed", although the details were not shared with the readers and "the resolutions were not generally favoured". Garcke eventually withdrew them, and presumably left with a black spot firmly fixed to the centre of his forehead! The bad smell of democracy had, not for the first time, permeated the committee room.

Meanwhile the B. B. K. A. were also having difficulty with the services being rendered by their secretary, E. H. Young, a paid servant of the Association, who appeared unable to carry out the duties because of serious illness allegedly caused by the increase in the amount of work he was having to do for the B. B. K. A. He also did similar work for other organisations, which presumably had not had the same detrimental effect on his health. Cowan, who had originally appointed Young, visited him and declared that he was in need of a complete rest. W. Herrod and another volunteered to do the work, with Young's brother superintending whilst the secretary recovered. Expenses and salary were to be paid to Herrod. Four months later Young was re-visited and found to be no better, they paid him off with a quarters salary, and made W. Herrod acting secretary. Herrod was instructed to write to the Youngs asking for the return of the Association's bookwork. The Youngs refused to co-operate. Cowan wrote several letters, all without the Council approval; Young resented the remarks therein. Retrospectively, the Council approved the letters, and asked the Chairman to write to the Youngs again asking for all the B. B. K. A. property to be returned, and informing them that they no longer required the address for their business. A month later, when the new premises they had been offered, proved inadequate, Cowan was asked to enquire of the Youngs, the cost of renting accommodation; - and they saw nothing wrong with this?

Cynical readers might be tempted to associate W. Broughton Carr's death, the affair at the Youngs, and the smart rise through the ranks of W. Herrod, but the evidence is purely circumstantial, and I would not wish to be implicated in such thoughts!

In February 1910 Cowan put W. Herrod in charge of the business affairs at the *B. B. J.,* and appointed him his 'editorial pupil'. In future the B. B. K. A. would be charged for advertisements in his magazines, and a little later when the *Journal* offices were moved to 23, Bedford Street, he also started charging the Association rent for accommodating Herrod. The successor had been confirmed; his Coronation would follow shortly. If there should be any doubt as to Cowan's choice of W. Herrod to succeed him, the following from his book *The Bee-keepers Guide* ...should dispel any doubts:

```
IN MEMORIAM
Clarum et venerabile nomen
Thomas William Cowan, F. L. S., F. G. S., F. R. M. S., F. E.
S., D.Sc., Ph. D.: honoured and respected in every country
where the honey-bee is cultivated; the doyen of English
bee-keeping, scientific, literary and practical; to whose
works, personal advice and assistance , the author hereby
reverently, affectionately, and gratefully acknowledges
his indebtedness for the apicultural knowledge acquired or
any success attained.
```

He has the Latin also –'illustrious and venerable name'.

Late 1909 the *Journal* began to carry a succession of letters on the reorganisation of the B. B. K. A. The letters appeared to have been selected on the basis that the proposals they contained were wildly impractical, or enable the editor to add a footnote pointing out the inaccuracy, as he saw it, of what the correspondent had written. Such was the case of a letter that included the claim that the Council was privately elected, a very sensitive point for the Chairman/editor. Cowan would never concede that the voting system, with the number of votes being dependent upon the money subscribed, resulted in the same individuals being elected every year. It was something he could never bring himself to accept. The temperature was raised with a letter from G. W. Avebury the Hon. Sec. of Cumberland B. K. A. headed: - *Has the B. B. K. A. Outlived its Usefulness?* In a footnote, the editor reminded his readers that the correspondent was entitled to attend Council meetings, but had never done so, and the same was true of most of the other County representatives, which he assumed indicated acceptance of what the B. B. K. A. were doing. A common tactic even today, and often accompanied by 'criticism is welcomed but only positive criticism'. Avebury was not prepared to accept the put-down and responded strongly, particularly in regard to the assumption made:

> We are satisfied thus far, that so long as the parent Association is content to move along in the same old groove it is useless to come all the way to London to help to do the usual routine. On one historic occasion only has the representative of Cumberland attended a meeting. Although a full number of the representatives the Editor refers to also attended, and almost unanimously laid a request before the Council for help in furthering what is universally held to be the most necessary reform for the advancement of bee-keeping in the counties, the result was such that the dissatisfaction then created remains still.

> Among the associations the position of the B.B.K.A. is unique, inasmuch as the only information, regarding its doings which reaches the public comes through the B.B.J. This paper, invaluable as it is to bee-keepers, has, though quite unintentionally, done the B.B.K.A. harm in the past. The B.B.K.A. is pointed to as the petted and spoiled child of the B.B.J. No mother defends a spoiled child from the slightest breath of adverse criticism so effectually, and anything not altogether laudatory of the B.B.K.A. is published under protest, with the result that a free public opinion of that body never finds expression. A fine feeling of regard for the Editor himself, mingled with a natural

aversion to be " hauled over the coals" in a critical
footnote from the editorial chair, overcomes the desire to
criticise from a sense of public duty.

Following his battle with G. Saunders the previous secretary for Cumberland, Cowan must have thought that one of the criteria for the secretarial position in that County, was the ability to ruffle the Chairman of the central Association!

The last Council meeting of 1909 had before them a proposed scheme for reorganisation – basically a B. B. K. A. for all Britain, comprising branches; the parent body would become a central branch; each County Association would become a County branch. Details of subscriptions and money flow were given, the full report to be published in Cowan's papers. Cost of the exercise was to be borne by Cowan and Garcke. Result, meeting with County representatives.

The first Council meeting of 1910 decided to pay Herrod £20 expenses and appoint him Secretary, at a salary of £60 per annum. The office of the B. B. K. A. was to be temporarily in those of the *B. B. J.* at 8, Henrietta Street, Covent Garden. Correspondence about reorganisation continued apace.

At the Council meeting on 17th February, Cowan made a statement reported thus:

He had worked for the Association, not with the idea of
being thanked for his services, but because he was anxious
to help forward the craft as much as he possibly could, and
was gratified to find that those services were appreciated.
He had acted as their Chairman for thirty-six years, and
he begged them to release him from those duties, as he now
felt that the work should be undertaken by someone younger.
At the same time, his services as adviser would still be at
their disposal. When abroad some years ago he tendered his
resignation, but in consideration of the pressure brought
to bear on him, and a statement by the late Baroness
Burdett-Coutts that his resignation would mean hers also,
he consented to continue while she lived, on condition that
the Council relieved him of as much work as possible. Her
death took place several years ago, and yet he was still,
in deference to their wishes, occupying the position; also,
through the unfortunate position of the Association during
the past year, the work and responsibility had been very
great. Fortunately, his health had remained very good; and
he had been able to get through it all right, though he
confessed it was a great tax upon his strength and time.
They would also have observed by recent letters in the

Bee Journal that some people *imagined* the Association had
suffered by being fostered too much by that paper. He was
very sorry if this were the case, but he assured them his
desire had always been to help them as much as he possibly
could.   The italics are mine.

He obviously thought this, and those he had made previously, were
resignations, but the statement above is not a resignation it is a request to be
'released from the duties'.   It is an excellent example of manipulation at its
best, one of his best and worst attributes.   All those who had asked him to
reconsider, knew well that he would not continue if the constitution of the B.
B. K. A. was significantly modified; it was a threat.   Cowan always said that he
was in favour of any change as long as it strengthened the central organisation.
However, there is no doubt that he was not pleased with the amount of work
with which he had recently been faced.   He was now seventy and thought
that he had retired to Somerset.   Much of the work was associated with his
business or need to be in control of the situation, and since Broughton Carr's
death, he had not been shielded from the continuing criticism that arrived daily
in the mail.

The conference was duly held in May with ten clauses to consider.   Fourteen
County Associations had expressed opposition to the scheme.   From a total
of thirty-four, two, Somerset and Devon were in favour.   The initial reported
remark from the Chairman, was that the B. B. K. A. had turned the corner
financially, so there was little point in discussing it.   Clauses one and two were
discussed and eventually two motions put and passed.   During consideration of
the next clause a motion was put and passed to refer clauses three to ten back
to the committee from whence it came for reconsideration.   That committee
was to be augmented by others.   Cowan does not appear to have taken part in
the discussion.   There was little enthusiasm for change by any of the Council,
who would ultimately have to agree to it, and other than a slight modification to
allow one member (to be approved by the Council), per affiliated Association,
to attend and vote at the A. G. M., there was little effective change.

The crisis had passed; the Council members and their leader could
contentedly sink back into their chairs.   Things would continue as before.   W.
Herrod had been their saviour; the books of the Association were now in order
and County Associations were being kept in touch.   For the first time the job
of secretary was being done properly, the Chairman had discovered a gem.
Herrod provided the bridge between the central body and the Counties that
had always been lacking.   Being from a working class background, he could
identify with the average County Association member and vice versa; he was a

worker, and of an age that gave him the energy to do the job and more, and he was not a Londoner. Herrod's appointment resulted in increased membership levels – the B. B. K. A. numbers doubled from 1910 to 1912. There had also been a grant from the Government of £850, with strings attached, but they did not appear to badly restrict the use of the money. When a member of the Council questioned certain expenses by Herrod, he was mocked and voted off the Committee. Not the first, nor the last time that this would happen.

On the bee front, in addition to foul brood disease, bee-keepers were having to face the greater threat of 'Isle of Wight' disease, first discovered in 1904. Initially, the B. B. K. A. had dismissed the discovery because it was not one of their experts that had found and reported it. When it was found on the mainland, it was too late to attempt confinement to the island. In the 19th edition of the *Guide,* in 1907, the author advised bee-keepers not to panic, adding that if it arrived here it would be unlikely to become an epidemic, which accurately reflected the attitude of the central members. By 1908 it was estimated that over 200,000 colonies had been destroyed. In the 20th edition of his book, Cowan changed his previous advice, saying that it was now causing great mortality and the only remedy was to burn bees and equipment; - time to panic – too late!

Cowan returned to his retirement in Somerset, only occasionally venturing up to London for meetings and to speak at conversaziones. He was keeping bees again and reported that in July 1912 he had a double swarm; not read the *Guide Book?* Cowan was in Switzerland in 1911 and in 1913 he went there for the final time to stay with his great friend Bertrand, who had ceased publication of his magazine *Review Internationale* nine years earlier, and was now fully retired. The gathering is shown below, a very disparate bunch. Cowan is seated left, but who is he looking at? Also present were Mr. and Mrs. Dadant from America on an European tour. It was a good decision not to leave it until the following year, when the festering political problems in Europe came to a head

Gathering at Bertrand's home in Switzerland.

with the shooting of the Archduke Franz Ferdinand and his wife by a Serbian terrorist in Sarajevo. Europe was at war; bee-keeping problems would soon seem insignificant.

Herrod informs us that Cowan had been sworn in as a special constable "and is taking his share of night work in guarding railway bridges and tunnels". Presumably his time in the West Kent Rifles fifty years earlier, proving useful. Herrod continued, imparting the information that Miss Cowan was a voluntary Red Cross nurse, and would no doubt go to the front. Alexander was to become the recruiting officer for Somerset with the rank of Captain. The youngest of the Cowan children, Percy, who had been on the editorial team of *Engineering* since 1905, joined the Army as a Civil Engineer, saw service in Mesopotamia, eventually rose to the rank of Lieutenant Colonel and was awarded the M. B. E.

The final conversazione of 1913 was on honey judging, and was notable for the attendance of C. N. White, who had fallen foul of Cowan because of an earlier bizarre event relating to foul brood. In the discussion following the paper, White said that he did not attend the B. B. K. A. meetings now because he had better things to do, and left immediately after his contribution, but the intended recipient of the remark was not there. However, the conversazione subject and an article by Joseph Herrod on experts, was the trigger for the Association dissidents to press their case for reorganization yet again. Correspondents were all saying basically the same thing – we are fed up with the old order, let us have a more democratic association that is far less dictatorial with transparent systems, and more answerable to members at all levels. It was obvious that a member of the B. B. K. A. committee putting himself forward for examination was far less likely to fail than someone from outside the circle; it was impossible to acknowledge that, if someone was helping to run the show, he was not competent, why else would he be there? It was a thinly veiled attack on W. Herrod who was 'running the show' and who had only recently, following criticism, become a first class expert. It was suggested that all those who became experts prior to that point would have to subject themselves to the new, independently carried out assessments. Not quite a revolution but sufficiently concerning that W. Herrod published several letters of praise for his work.

On a lighter note, J. C. Bee Mason, who was later to become a war photographer, and official photographer on Shackleton's final voyage to the Antarctic, had made some movies on bees. In February 1913 he showed them at the London Opera House for three weeks, where he appeared on stage with a band. Later he took the show on tour, adding to the films he had made. It was not surprising that the top of the B. B. K. A. were not pleased – "mere trivia….. passing fancy….. will not last". Later he showed them at a conversazione, where B. B. K. A. Council members tried to 'talk him out', but the assembled body stayed on to see the films and demanded that he return, and next time earlier in the running order.

In the December, 1914 issue of the *Journal,* W. Herrod, who presumably was not doing his 'bit' for the country, published an article taken from the *American Bee Journal* written by C. P. Dadant about his travels in Europe. It contained little of interest, but Herrod indicated that, had Cowan not been busy defending the country, he would have taken the blue pencil and scissors to the article, removing the parts complimentary to him, not wishing to have praise made public. There were very few complimentary passages, and nothing that had not been in the *Journal* before. It appears to indicate that the Senior Editor was no longer actively involved. It is simply adulation, but the dose was repeated the following year.

Cowan portrait from the collection of the
Somerset Archeaological and Natural History Society. Age - mid seventies.

The next year started with the following: to use a *B.B.J.* 'ism' - any comment would be superfluous:

```
NOTICE.
Owing to the dislike of one of my ancestors to a long
signature he dropped a portion of his name. For several
generations this caused no inconvenience, as they did very
little business. Owing to the continued increase in my
business interests, the dual situation of one signature
for ordinary use and another for legal matters has become
impossible; therefore, on and after January 1st, 1915, I
shall assume my full name of W. Herrod-Hempsall. I shall
be grateful if all correspondents will kindly note this
-and address me, also make out all cheques or documents and
insert in any list my name, as above,
W. Herrod-Hempsall,
hitherto commonly known as W. Herrod.
```

So brothers William and Joseph were now hyphenated. Cowan, once again, asked to be released as Chairman of the B. B. K. A.. He was requested to continue, and acquiesced. He was made President of Somerset B. K. A. Not unnaturally, at seventy-five years of age he did not venture to London for any Council meetings. Mrs. Cowan again seriously ill at the time of the next A. G. M. Joseph Herrod-Hempsall became manager of the magazines taking over from William. So with the Herrod-Hempsall brothers effectively running the *B. B. J.* and *Record*, and William becoming increasingly powerful in the B. B. K. A., bee-keeping had slipped seamlessly from the autocratic control of T. W. Cowan to that of the Herrod-Hempsalls.

In 1918 Thomas and Fanny made their final move, this time the relatively short distance to Sutherland House, Clevedon, as always Edith going with them. Alexander and Frances moved to Milverton, naming their house 'Loomis'. Alexander was still at this time running his tyre business in Taunton. At the same time Cowan sold the proprietary rights of his magazines to William Herrod-Hempsall, who in turn sold them to his brother two years later following his appointment to a full time position with the Ministry of Agriculture. Cowan's name was retained as editor, but it was acknowledged that his capacity was more in the nature of adviser. He remained Chairman of the B. B. K. A. Council.

However, there were 'clouds gathering'. In 1919 by Dr. Ahmed Z. Abushady, an Egyptian scientist and poet, formed the International Association of Apiarists – The Apis Club. The club immediately commenced publication of their magazine *The Bee World*. In the same year *Bee Craft*, initially the magazine of the Kent B. K. A. appeared, and became another voice for the disenfranchised B. B. K. A. members. The future was to be no more settled that the past, but that is the subject of someone else's book.

At the Council meeting of November 17th, 1921 it was decided to present T. W. Cowan with an illuminated address in recognition of his long and dedicated service to the Association. It was necessary to subscribe to get your name on it. The presentation would take place at the next A. G. M. after the Council elections and the announcement that the Master of the Worshipful Company of Wax Chandlers would be raised to Patron from President. Cowan would then be elected President.

Unusually Cowan attended the A. G. M. His election was made "with acclimation". He responded by saying that he had no idea that it was contemplated; the request to become President was spontaneous and had been received so enthusiastically that he could not do otherwise than accept (long and loud applause). What did he think was going to happen? The question obviously arises, – why did it take them so long to arrive at the obvious solution? After tea, and presentation of the illuminated address, books, and a silver fruit dish for Fanny, Cowan responded with a carefully prepared speech - the last time he would address 'his' Association members.

Presentation to T. W. Cowan upon his elevation from Chairman to President

In 1923 the Apis Club, of which he was a Vice president, presented Thomas with its Gold Medal even though he had many disagreements with Abushady since the establishment of the club. A year later the Gold Medalist wrote to *The Bee World* objecting that the story about Hoge, an American, being the purveyor of honey to the Queen in 1884, was allowed to pass without correction by the editor. He quoted the whole letter that he had received from the Lord Steward, stating that Hoge held no such appointment. He rightly termed Hoge "an impudent imposter", but the fact that he was still trying to effect closure on this matter forty years after the event, indicates his continued annoyance that the message had still not been accepted.

Over the coming years the venerable leader would, receive severe criticism at the hands of *The Bee World,* demonstrating its independent nature. The first indication that the editor was to allow it, appeared a few months after Cowan's letter, when a correspondent complained that the editor had taken a charitable view of that correspondents earlier letter on 'Guide-Book Mathematics'. His final paragraph read:

> One only has to refer to Mr. Cowan's *Honey Bee* to realize how little anyone dares to criticise a recognized authority. On p. 120 (second edition) the author refers to a queen as weighing 100 grammes – more than a very large hen's egg! Obviously this was a clerical or printer's error; but the fact that no one had to question it in the first. A scientific writer actually depends on criticism from readers in order to enable him to perfect his work; and it is distinctly unfair on him when a gang of dogmatists forbids all criticism. I take it that the reason why your paper was so unpopular among certain classes at its inception was because, being run on scientific lines, it encouraged criticism as a means of arriving at the truth.

There would be more in the future.

Cowan spent much of the last two years of his life writing the history of the first fifty years of the B. B. K. A. as he saw it. It is not an objective or, in parts, accurate account, but it was eventually published as a book, helped by a donation from it's author. Perhaps because he was not able to get about so well in his later years, he occasionally entertained bee-keepers at his house in Clevedon, and the photograph showing Mr. and Mrs. Cowan with daughter Edith in their lounge was probably taken on one such occasion. This photograph was discovered by David Charles and is taken from his splendid *Somerset Beekeepers and Beekeeping Associations,* published in 2005. The saying 'a picture paints a thousand words', could never be more applicable than here. It is worth more than a glance.

Sutherland House

Thomas, Fanny and Edith, Sutherland House, c. 1923

Annie Betts, editor of *The Bee World,* was one such visitor. She wrote in letters of her visit there, that Cowan was very accommodating in loaning her books from his library and discussing microscopic techniques with her, but she felt that this was because he did not consider her a threat. She also said that she got the impression that he treated his family with contempt. The friendship given by Thomas did not stop Annie writing a scathing review of the latest edition of the *Guide* in 1924, or an editorial on the Digges affair headed *A Fallen Leader.*

As would be expected, W. Herrod-Hempsall visited and persuaded Thomas to pose with him beside his car. Thomas looks very frail, and I believe this to be very close to the end of his life.

Thomas with W. Herod-Hempsall, mid 1920

Another set of visitors were entertained in December 1925; it was to have been a visit by Dr. C. R. Killick, and W. H. Ashwin, allegedly to see the famous microscope. Both were prominent members of the Somerset B. K. A. Cowan, believing that the microscope was just a guise for introducing the topic of reorganization of the central association, insisted that L. Snelgrove and R. Beck, also members of the local Association be included. In his article for the *B. B. J.,* Killick described the event: inspection of microscope, tea and inspection of books in library. It was a non-event. However, Killick, said that he had tried to interest Cowan in the microscopical section of the Natural History Society of the Archaeological Society of Somerset, but he had declined; a strange decision, given his alleged interest. When Beck took the microscope out of its case, he said, "...this microscope was made entirely by Mr. Cowan from castings prepared by my father". Reference to the *English Mechanic and World of Science,* which was very strong on microscopes, reveals that the firm of R. & J. Beck, as well as selling microscopes ready assembled, also sold kits to enable users to assemble them for their own specific requirements. The impression always given was that Cowan had made it with his own hands from scratch.

Killick finished his article thus:

> If at any time the original edifice of the B. B. K. A. should
> be pulled down for alterations, let us mark all the stones
> that may be still fit for service in the new structure, and
> let posterity be told that these stones once stood in the
> in the first Bee-keeping Association of this Country. If we
> can build better than Mr. Cowan and his associates built in
> their time, we, too, shall earn part of his reward.

Any doubts as to whether he had made the correct decision by insisting upon two allies being present at the meeting, would have been dissipated two months later when at the Somerset B. K. A., A. G. M., Killick proposed and Ashwin seconded, that the central Association should be reconstructed. A month later reorganisation, was yet again 'kicked into the long grass' by the Council, stating they were "tired of this hardy annual", but the edifice was already under attack.

A few days after Killick's visit, Cowan let it be known that he would not be able to receive any further visitors because he was too ill. Mid February the President wrote to the Council:

> Dear Mr. Reid, - I have received the resolution of thanks
> passed at the last Council meeting, and feel deeply grateful
> for the kind feeling therein expressed, both to Mrs. Cowan
> and myself.
>
> The compilation of the history of the Association, with
> which I have been connected since its beginning had taken
> considerable time, but it has been a labour of love, and I
> am quite pleased that the Council has considered it worthy
> of its appreciation.
>
> With best wishes for the prosperity of the Association, I
> remain, yours faithfully,

In his 1925 seasonal greetings to the editor of the *American Bee Journal,* Cowan, rather prophetically wrote:

> I thought that I would let you and Mrs Dadant know that

at 86 I am still able to enjoy reading the American Bee
Journal, which I receive very regularly and so keep in
touch with what is going on in your country. However, I am
not able to do very much as I have been laid up with what
the doctor calls influenza, but I think the doctors give
this name to any illness they know very little about.

These were his last two recorded communications; Thomas William Cowan died at home, in Sutherland House, Clevedon, on the 23rd May 1926, aged eighty-six. Both his parents had died at the same age.

The local paper carried an obituary that included "he caught the idea of wireless by installing trumpet-like instruments in each corner of his dining room". I was unable to find any other reference to this, even though my curiosity was aroused.

Herrod-Hempsall in his notice of Cowan's death linked it to a fall he had on the 17th, but there is no such information on his death certificate. Even death had to be heroic for his chosen one. The obituary in the same magazine was surprisingly short, and followed by an even briefer account of his funeral on the 26th, which was also reported in the local newspaper and most other bee-keeping magazines of the time. The mourners did not include Fanny, who presumably would have been too frail to attend, or Edith who would have stayed with her. Rather surprisingly, there appeared to be no representation from the B. B. K. A., although several from the local Association were present. Neither of the Herrod-Hempsall brothers were reported as being present.

Headstone of Thomas, Fanny and Edith

Chapter XIII

In Conclusion

Thomas William Cowan was a very, very lucky man; born into a middle class family, received an excellent education, experienced a first-rate apprenticeship, and married the only daughter of a wealthy businessman. He was a complex character, who fiercely guarded his private life, resulting in him having few close friends and making it very difficult for the historian. A cursory inspection reveals that he did things very correctly, as he saw it. W. Broughton Carr and the Herrod-Hempsall brothers claimed that he shunned publicity, all mentioning the blue pencil and scissors of their senior editor. Both these claims are only partially true. Col. Walker, who claimed to be a close friend - he was certainly an ally, wrote that Cowan was 'reserved', but to his closest friends he was 'genial', not an overwhelming endorsement of friendship. As he grew older he became more aloof, and gave the impression that he was superior in class to those he met in beekeeping. The only exceptions being those with Sir or Hon. preceding their name, Dr. or Rev. definitely not included.

R. O. B. Manley related the apocryphal story, that when Cowan arrived late for a B. B. K. A. talk, as he entered, all those present stood and only resumed their seats after the Chairman had taken his. The speaker then proceeded to start again at the beginning of his lecture. It was no more than Cowan expected, but any 'reserved' individual would have been deeply embarrassed. Manley would, along with Annie Betts, later become the nemesis of W. Herrod-Hempsall and Cowan.

He was always the one to interface with Royalty, Baroness Burdett Coutts, and Government personnel. His closest friend was Eduard Bertrand, his equivalent in Switzerland, whom he visited many times and entertained at his houses in Britain - a rare event.

He did not smoke, was teetotal, an avid letter writer and lauded as a great multi-linguist. It is not possible to confirm the latter, but he did none of the translations of his works, and his lengthy holidays in Europe and school in Russia would have helped. The Latin he used in his articles was only 'sayings' to be found in any good dictionary. He described himself as 'warmly religious', and as Cowan never used irony, we can take that at face value. He was a member of the Church of England, and was involved in some church based charitable activities. He loved travelling; his wife's money providing the leisure time and requisite finance, the advent of the railways and steam ship providing the necessary means. However it was alleged that he treated his family with contempt, of which there is some evidence.

He was a serial business failure, and had it not been for Fanny's inheritance, I believe his bee papers would have ended in the same manner as Kent Iron Works, Trefanynt mine, British Honey Co., and Simmins bee businesses.

His authorship skills were very strong. He wrote in a very clear, concise manner, the ideal style for text–books. Of their time I believe his three major works, *The British Beekeepers Guide Book, The Honey Bee,* and *Waxcraft,* were excellent, although none were fault free. The same adjectives can be applied to his leaflets and magazine contributions. He was particularly strong on bee management although often not detailing his experimentation, just giving the conclusion. However he never managed to communicate well with the 'cottager' bee-keeper, and certainly never associated with them. Rich and poor were kept apart in Victorian England, aristocratic power and accumulated wealth remained largely undisturbed and Cowan had no intention of upsetting the status quo, even though he was of a lower class. He never acknowledged his mistakes.

Cowan used Latin to appear scholarly and as a sign of his class. Many believe that further or higher education might have cramped their style, and this might have been true of Thomas, but he clearly wanted others to believe that he had it. There is little doubt that Victorian scientists and Engineers had difficulty getting accepted compared to their counterparts in Europe; for example, Germany recognised Science as an academic subject by the mid nineteenth century. Natural Science was viewed even less favourably; in a debate on the Public Schools Act in 1868, it was said that Natural Sciences do not provide a basis for education. Aristocratic members of the government were afraid that it would provide the means that, say, sons of small businessmen, to succeed at the expense of Army officers, 'who had learned truth and honour at home'. Yet an over-riding memory of the Victorian age is the major engineering achievements.

Cowan unashamedly used his magazines to prostitute his publications, even though he abhorred this in others. Queries were often answered with a reference to 'chapter and verse' in one of books or papers, and a sure way to get your contribution published was for it to contain mention of one or more of his books. This was essential to be in 'Homes of the Honey-Bee'. He used the editorial section in his magazines to promote his views and often refused to publish the alternatives, or he would publish the views of the opposition, in order that he could mock it in a footnote. His magazines only contained positive reviews of his works, often written by his allies or friends. He had total control.

He was generous. Over the period of his membership, he contributed to all calls for money; donated equipment and trophies as prizes and gave many books to the B. B. K. A. library. However, he did not bequeath his library to the Association resulting in it having to purchase the collection from Benefactors of his estate. Mainly he gave of his time, energy and intellect, supporting his belief that the B. B. K. A. was, and could only be, a philanthropic organisation, and of course, preferably led by himself.

Without his leadership I have little doubt that the B. B. K. A. would not have had the limited success that it did enjoy, or have even survived at times. When he was not active the Association was moribund. He instigated the examinations and awards system, but he was also responsible for giving examination awards to his supporters. However, he refused to recognise the need for change in the B. B. K. A. even though he offered on many occasions to resign from Chairmanship of the Council. He used his position, often, to champion his personal agenda and when it suited him, by-passed the correct channels. Despite much critical correspondence he failed to grasp the need to 'clean up' judging and organisation of the Association shows until it became an embarrassment – even for him. He was known in most of the bee-keeping countries of the world, which was advantageous to the craft in Britain. However, he continued in post even when he was not capable of carrying out the work; something he decried in others.

His most notable failure was his inability to achieve bee disease legislation, followed closely by his blocking of any move towards 'his' Association becoming democratic, or a meritocracy, blackmail coming naturally to him.

He possessed an inventive mind, and applied it to bee-keeping equipment, but inevitably there was the good, the bad, and the downright ugly.

He was vindictive, arrogant, hypocritical, selfish, conceited, and fiercely patriotic. He did not possess the ability to concede, believing that the label 'Gentleman' entitled him to be correct at all times. He made instant judgments of others on the basis of class, colour, or nationality. The manner in which he dealt with Samson, White, Hewitt and Digges was disgraceful; his description of the indigenous population of Africa appalling; and his belief that no foreigner could possibly have invented something before the British in general and himself in particular, unrealistic. To quote Sydney Smith again "Never try to reason the prejudice out of a man. It was not reasoned into him and cannot be reasoned out of him".

However, it is his 'being economical with the truth', to use the popular euphemism, that is the most worrying personal characteristic, and cannot be excused on the basis of period, as some of the others might be.

It is true that the Victorians lived with double standards both individually and collectively which makes them appear humbugs and hypocrites, and the

Victorian establishment was quick to close ranks against criticism. Parallels can be drawn here with Cowan and the B. B. K. A. There is also a very strong comparison to be drawn between the B. B. K. A. and the British Army of Victoria's reign, where commissions were initially purchased, and when that ceased, the officers continued to select the officer class. It was what occurred at the time.

The Victorians witnessed a cholera epidemic in 1866, brought about by their laissez faire attitude, which appeared to prefer death to state interference. The latter would, of course, require funding from the state, and for a class that considered the level of income tax in 1875 of 4d in the pound to be punitive, this was not viewed with any enthusiasm. Whilst Cowan and the B. B. K. A. canvassed for taxpayer's money to be spent on bee-keeping, they resented paying taxes and any intervention by the State in the manner that the grant was spent. I think that I have already mentioned hypocrisy. Another parallel can be drawn between successive Victorian Governments' lack of enthusiasm to widen the franchise, and the insistence by Cowan and his friends that the rules for election of Council officers remain unchanged.

Many of the Victorian views, appear so unenlightened to us in the twenty-first century. The system of hierarchy very prevalent in the Victorian era is now considered archaic, even though remnants still prevail; patronage and nepotism has generally been supplanted by a meritocracy – even in bee-keeping. It is this, which makes any genuine assessment of a Victorian individual difficult. A visit to a present day living museum might provide a glimpse into Victorian life - the prevailing stench in the streets; the lack of personal hygiene, - not just by the servants; the difficulty and time taken to travel any distance, - although the trains were on time; the lack of services, the attitude to religion, socialism, Darwin, death, etc. etc.

With advancing years we tend to look back, posing many 'what ifs'. Reading his Jubilee history, I do not believe this is something Cowan indulged in. However, what if a bunch of incompetent, aristocratic army leaders had not fancied one last hoorah, resulting in the British government declaring war on Russia in March 1854; what if Thomas and Fanny had not met and married; what if Henry Michell had lived longer or left his money differently; what if Tegetmeir had chosen bees rather than poultry; what if, what if, what if, …..- the slender thread whereby hangs this biography.

Thomas William Cowan was a lucky man, whether British bee-keeping was lucky to have him in control for fifty years, cannot be accurately assessed because there is nothing to compare with. However, for good or bad, he had a major influence on the way we keep honey bees today.

Appendix I

The English Mechanic and World of Science.

T. W. Cowan's first entry was a short item in response to a query on how to varnish engravings. There followed short contributions on mastic varnish, browning a gun barrel, a mixture for storm glasses, how to colour Staffordshire bricks, where to purchase Zoetrope slides, and recommendation of a book on how to prepare lectures. Two issues later there followed the first of many of his entries in the exchange column, free of charge for subscribers at this time:

```
Yacht engine.  I will exchange new yacht double cylinder
trunk engine of about 5 h. p. or 6 h. p. for a first class
microscope.  J. (T) Cowan, Beckenham, Kent.  [October 30,
1868]
```

This was the type of engine that he made at his factory 'Kent Iron Works'. He was using *The English Mechanic* to help shift some product, without the cost of advertising! The same month carried his first lengthy article on how to make a Ruhmkorff coil. He appears to have made one himself. It was a spark coil energised from a direct current supply. Over time this drew much criticism from those that had tried to make one using the Cowan instructions and failed. His response was the same as would become familiar to beekeepers in the future – robust defence; wrong size wire, wire broken, insulation not dried adequately etc. The advertisement above reappeared, and he continued to demonstrate the breadth of his scientific knowledge with pieces on where to purchase wood carvers tools, (he practised wood carving until his death), keeping moths from drawers – lumps of camphor, and how to solder copper, which again drew many reports of failure from others, with a phrase that Cowan would be confronted with on many future occasions "It did not work for me" only to be told that it must be the way you are doing it. It was the usual 'discussion' that *English Mechanic* was well known for.

Cowan continued to be active in the small advertisement section:

```
PRINTING MACHINE.  I have a foolscap size printing machine,
a large quantity of plain and ornamental type, figures,
border, brass rule and furniture, inking roller, composing
stick, compositors' frame and rack and a double letter
case.  I will exchange the whole for one of Cunningham's
ornamental wood cutting machines {carving}.  T W Cowan,
Beckenham, Kent.
```

Could this have been the machine he used to print the leaflets advertising his products from his factory?

> For sale. A pair of 4 h. p. yacht trunk engines complete
> except reversing gear, quite new. Will be sold cheap £30.
> Apply Mr Cowan, Beckenham, Kent. [January 1869].

The selling of trunk engines, the basis of his engineering business, from his home address indicates the closure of Kent Iron Works to have been late 1868.

The *E. M.* entries continued – how to amalgamate zinc battery plates, more on the Zoetrope, and response to a suggestion by another subscriber for the use of a steam engine to drive a clay-crushing machine in brick works. In this he indicated that he had recently been to Germany. Brick making was one of his father-in-laws businesses.

There followed a longer article by Cowan concerning the decision on the type and source of stone to be used on the Thames Embankment included elsewhere.

The entries continued in the exchange and for sale columns:

> Organ Acordian. I have an organ accordion with one stop,
> with pedal action and stand. I will exchange for dissolving-
> view lantern or sewing machine. T. W. C. Beckenham.

> Exchange Trombone. I have a German trombone with four
> resolving valves by Fetche & Sons of Berlin. I will exchange
> it for a good microscope with 1" and ¼" objectives. T. W.
> C. Beckenham.

> For sale. A first-class single-barrel SPORTING GUN by Hurst
> in polished oak case complete. £4 10s. Apply to Mr Cowan,
> Beckenham.

> DRAWINGS. Working drawings of a 20" cylinder pumping
> engine and other drawings for dissolving view slides or a
> sewing machine. Mr Cowan, Beckenham.

> Electrical Machine. A cylinder electrical machine and other
> articles for the first five volumes of English Mechanic. Mr
> Cowan, Beckenham.

> Exchange. For scientific books or articles a glass case of
> butterflies, a pair of floor skates and other articles will
> be given. T Cowan, Beckenham, Kent.

> VELOCIPEDE. A working drawing to scale, with every detail

separate, of the best French two-wheel Velocipede.   Post
free 24 stamps.   T Cowan, Beckenham.

The latter advertisement recurred during the next few months. All the above entries were in the first four months of 1869. There were also responses/articles on recipe for coloured ink; plant identification; a simple method for wrought and cast iron and during one exchange he revealed that he possessed a copy of "The Philosophy of Electrical Psychology" by John Bovee Dods.

He gave details of his Steam Carriage for road, given elsewhere.

There were a couple of follow-ups to this item, one correspondent asking Cowan where the boiler of his carriage could be obtained and the price. Cowan responded by asking for his name. The information was not published in the magazine. The second response was in the notices to correspondents:

THOMAS W COWAN.   We will forward any stamped letter you
may enclose to us to A. B. C.   We do not think that our
correspondent has yet constructed one of his proposed steam
engines.

This was typical of the bizarre, understandable exchanges that took place in the magazine.

There was further banter on his induction coil construction, and several correspondents requesting that Cowan redeem his promise and inform on the use of the coil. He never did. He became involved in a discussion on refrigeration recommending to another correspondent a specific make of refrigerator where the liquid to be cooled ran over a series of tubes carrying cold water. Older rural readers will be reminded of a milk cooler of the 1940/50s in every farm dairy.   Could this be the origin for his 'honey evaporator'?

He continued to be a frequent contributor to the personal advertisements:

Exchange; a pair of new skates for fifth volume of "Our
Mechanic".

A previous entry partly successful.

For sale.   Letter copying press and stand, composing
frame, large aquarium, German four valved Trombone, parlor
croquet, retriever dog, kennel, beehives etc.   Price list,
T W Cowan Beckenham.

This has all the appearances of a clearout prior to his move to Horsham. It was mid July 1869. No takers yet for the printing press, trombone or dog and his abode!

> Exchange. Hydropult and other articles for books or
> scientific instruments.

This was a water pump, possibly for the garden, as shown in one of his patents.

> Exchange. Books or scientific articles for a stereoscopic
> camera.

> Exchange. Scientific and other articles will be given for
> a gazogene.

A gazogene was a device for making Soda water or aerated liquids on a small scale. It was two interlinking spheres water in the bottom one and tartaric acid and sodium bicarbonate (produces carbon dioxide) in the other. This was similar to the kind of device that he used to produce his aerated honey for showing as mead.

He corresponded on recipe for waterproofing fishing lines; how to destroy ants; (I could join the band of 'it did not work for me', here); the mix for a brick; how he stupefies insects. He decried the state of education in the Country, quoting an article from the *Rugby Meteor* which showed "a lamentable state of ignorance" quoting much of the piece to prove the point. In a footnote the editor pointed out that the English was so bad it was obviously a spoof from one of the brightest students there!

He contributed an article on alleviating vibration of bicycle handles. It involved a strong door spring fixed to the iron hook in the centre of the footrest and from each of the eyes a piece of bright copper wire is connected to the handles and "is very ornamental". A bicycle manufacturer had seen it and been so impressed that he now uses it on all the bicycles he sells.

He continues to have to defend his induction coil, writes on parasites in pork, and claims that a published suggestion for preventing seasickness will not work. This latter entry was signed T. W. Cowan C. E. Ph. D. There is no listed abbreviation C. E. but it could have been one of the many misprints that characterised the magazine's columns, or he used it believing it was for Civil Engineer. It was in the April 11th 1870, edition and the first time that I found where he used Ph. D.

The attempted selling continued.

> Exchange. A beautiful trombone for a sewing machine.

> Exchange. For a microscope, scientific or other articles
> will be given.

Black trained retriever dog and kennels, books, skates and a number of articles for books or scientific articles.

For sale. Well trained black retriever dog £3 value ten guineas, or for exchange.

For sale. Square pianoforte by Broadwood £10, black retriever dog £3.

For sale. Good "Broadwoods" piano, large retriever dog, parlour croquet board and other articles.

Exchange. A black retriever dog and kennel for microscope or articles of equal value. T W Cowan, Beckenham, Kent.

This was his last entry from Beckenham. It was August 26th 1870. Despite the apparently excellent value, the unwanted animal was proving difficult to re-home! Reference to Henry Michells diaries shows that he did not take to the shooting lifestyle, probably the impetus for selling gun and dog. No longer attempting to be part of the 'Country set'.

Having moved to Horsham, his next entry was not until March 29th 1872.

For sale. Having more vegetable seeds than I require for my own use I shall be glad to exchange for plants or dispose of the surplus at 4s for the collection of twelve packets. Melon seed 6d the packet. T W Cowan, Horsham.

For sale. A few duplicates of mounted microscopic objects for exchange or sale.

For sale. Microscopic objects – Dendritic spot on paper and other objects.

It would appear that the microscope had been acquired and the dog and home disposed of in the deal. On 3rd February 1875, Thomas was elected Fellow of the Royal Microscopical Society.

The entries from Thomas were now very infrequent and his last appears to have been on 9th January 1874, when he reported on an exhibition of fossils in Horsham.

Conclusions.

He claimed to hold the doctorate qualification a year earlier than the revelation in the 1871 census. In all other entries where he used a title in the personal columns it was 'Mr'.

The personal advertisements selling non -domestic itcms suggest that he had ceased his connection with Kent Iron Works by 1869. The entries in Henry Michell's diaries indicate that he went to work with Michell's son in the brewery business. This explains the desirability to move to Horsham. But it also coincided with the birth of their first child. His selling off of domestic items would indicate either a shortage of money or space in the new Horsham house.

There was no description of his hive in either 1865 or 1866 *English Mechanic* despite claims ad nausea to the contrary. The *E. M.* did not carry beekeeping content at this early date. Cowan contributed no beekeeping articles to the *E. M.* The beekeeping content when it did commence (1869) was mostly by C. N. Abbott - practising for the *B. B. J.*

Appendix 2.

B. B. K. A. Committee Bye-laws.

160   THE BRITISH BEE JOURNAL.   [Jan. 1, 1875.

BRITISH BEE-KEEPERS' ASSOCIATION.

THE Committee met by adjournment at Beaufort Buildings, Strand, on the 8th ultimo. There were present— T. W. Cowan, Esq. (Chairman), Messrs. Hooker, W. Abbott, Atlee, Cheshire, Hunter (Hon. Secretary), and C. N. Abbott, (Treasurer). The *business* of the Meeting was the discussion of a Set of Rules, brought forward by the Chairman, for the guidance of the Committee, and much time was usefully spent in their consideration. Some alterations from the original were made, and the result, proposed by Mr. Atlee, seconded by Mr. Hooker, and carried unanimously, was as follows:—

COMMITTEE BYE-LAWS.

1.—All persons whom it is proposed to add to the Committee as Members shall be nominated at one meeting, and proposed, seconded, and voted for at the following meeting.

2.—Three shall form a quorum.

3.—The order of business at meetings shall be as follows:
1. Minutes of previous meeting.
2. Secretary's report.
3. Notices of motion.
4. Reports of Sub-Committees (if any).
5. Other business.

But the Chairman shall at any time be at liberty to vary the order of subjects upon the Agenda paper.

4.—Motions may be brought forward by a Sub-Committee or by the Hon. Sec. without notice, but other members shall give notice, in writing, at previous meeting, of any motion which they propose to submit, provided that motions of which no notice has been given may, if no objection be made, be at once brought under discussion, but shall not be put to the vote until the following meeting, if any member of the Committee object.

5.—The Chairman, if called upon, shall rule that in discussion of resolutions each Member shall only have the right to speak once to each separate question, the proposer having the right to reply.

6.—The Hon. Sec. shall be an *ex officio* Member of all Sub-Committees.

7.—One or more Auditors shall be appointed to audit the accounts each year, either by the Members at the General Meeting, or, failing such appointment, by the Committee, and that the accounts be made up to 31st December in each year.

8.—It shall be the duty of the Chairman to record in the Agenda Book all orders passed by the Committee.

9.—It shall be the duty of the Hon. Sec. to direct and supervise the paid officers; to obtain estimates for all contemplated expenditure; and generally to prepare the business for the Committee, to see that their instructions are recorded and carried into effect, and to call all meetings.

Appendix 3

Some Financial Considerations

To arrive at an accurate assessment of T. W. Cowan's finances is impossible. However, as part of the overall picture I felt it necessary to make some attempt.

Between 1860 and 1910, official statistics show very little change in the value of money. Using the base year, 2004, the 'present worth' factor for comparison is just over sixty.

Income.
1861 Gifted Kent Iron Works.
1861 – 1868. The income from his engineering works is unknown because it was not a registered company and there are no public accounts. The number of servants following his marriage was normal for a family with an annual income of approximately £300 p. a.
1865 Wedding day gift of £1,000.
1869 – 1874. Paid by Henry Michell for working in his business. No hard evidence, but Henry's bequest to Thomas indicates it was probably £300 p. a.
1874. Fanny inherits. Study of the financial literature covering the next fifty-two years, indicates a return of between 3% and 4% probable, giving an income from her trust fund of £750 - £1,000 p. a.
1874 Thomas inherits £300 p. a. until 1886 from Henry Michell.
1886 Thomas inherits estate, value approximately £2,000, from his parents and is paid expenses of over £100. His inheritance here would eventually rise to £3,000.
1890 Profit from sale of Comptons Lea not likely to be greater than £400.
1887 – 1918. Proprietor of *B. B. J.* (1887), and *Record* (1890). I do not believe he paid for either of these titles.
1918 Sold propriety rights of magazines to W. Herrod-Hempsall. No figure available.
1881 – 1926. Income from publications.

Despite considerable research, I found it impossible to assign a reliable figure to the monetary gains from his publishing business. However, I calculated/ estimated that in the forty-five years up to his death, Thomas made £10,400, but the statistical tolerance was large.

Putting these numbers into context here are some figures from the late 1800s.

Agricultural worker earned £30 - £40 p. a. depending upon job and locatio

Skilled factory worker was paid £62 p. a.

Live-in servant paid £12 -£23 p. a. depending upon position

Trollope informs that a living paid £900 p. a., which gave a comfortable existence.

A house in Bloomsbury, London, could be rented for £120 p. a.

To man the bee tent at a show a B. B. K. A. expert charged 10/6 per day plus expenses,

Isle of Man bed & breakfast 3/6, full board 5/- per day

And for a few prices that I know will probably end in your committing this book to the fire:

Honey jars; 17/- per gross; jelly glass type 10/6 per gross; corks 2/- per gross.

B. B. J. bound copy 10/- for year.

I. B. J. 3d monthly, 4/- annually including post.

Sections' 2/6 per 100.

Beeswax, 1/3 – 1/ 4 per lb.

Sugar pre 1914, 1 ½ d per lb, after 1918, 9 ½ d per lb.

Zinc queen excluders, 7d.

Honey knife, 2/3.

Smoker, 2/6.

Veil, 2/6, "or kit for the lady to make one", 1/6.

Cowan extractors, Amateur, 30/-; Rapid, 35/-; Automatic, 50/-.

Cowan hive with 11 frames, 30/-, with 13 frames 35/-.

Cottage hive complete 35/-

Bee house, £11. Few takers it appears.

Swarm with common black queen, 15/- - 20/- weight dependent.

Black queen 2/6.

Swarm with Italian queen, 30/- - 40/-. Italian queen 5/6.

Driven bees 1/- per lb.

Nucleus 12/-.

Honey, 10d – 1/- per pound section; 10d – 1/- per lb.

Heather honey, 1/- - 1/6 per lb. These honey prices were down on earlier years.

Appendix 4

Hewitt letter to Baroness Burdett-Coutts

B.B.K.A.

Cambridge street, Sheffield,
April, 1898.

To the Hon. the Baroness Burdett Coutts.

Madam,—As you either fail or refuse to comply with my demands of February 18th, 1892, and September 2nd, 1897, to have the truth of certain charges I make against Messrs. Cowan and Carr and other prominent officers of the British Bee Keepers' Association, of which you have been the president for the past 20 years, investigated, I am now reluctantly compelled to assume that you are quite cognisant of and a party to all their nefarious proceedings, or you would most certainly have had them seen into otherwise, which is the least the president of any society with the slightest claim to honour and common honesty can do.

As I am determined that such vile conduct shall not escape punishment, as well as all who have been aiders and abetters, I am driven, through your refusal to investigate my demands, to address you in public by means of an open letter, instead of private ones as heretofore.

Ever since I can remember, the public Press has been continually announcing some noble or philanthrophic deed of yours, until your name has become a household word as synonimous of everything which is pure and good; and, while I have for years lamented your conduct, miscalled "generosity," "philanthropy," etc., I never once doubted your motives, no matter how mistaken I regarded them, until you wrote the notorious letter to the painters in September, 1894. I did not nor do I even now approve of those painters' demands, but they were at least consistent, whilst your reply to them was the most inconsistent of anything I have ever read. You speak grandiloquently of the "stablemen" being "ready and willing to add something to their wages, after their ordinary duties were done, so laying by a little for a rainy day, instead of spending their time in idleness, and gaining some useful knowledge which would be of service in after life, should they exercise their undoubted right to improve or change their position."

These are the words you use, yet you penned them whilst in possession of my letter complaining and demanding an investigation against a society of which you were and are the head—a society which was being used, with your name, to prevent me doing lawful and beneficial work—a work the aforesaid society claimed to be doing its utmost to accomplish, and that the one who could accomplish it would be one of the greatest benefactors to bee keeping. These two facts are utterly at variance with each other. The painters wanted only men belonging to their society to be employed in painting. The society of which you are the president decided, because I was not "one of them," I was to be crushed by every means in its power, and they even went so far as to decide that Mr. Cowan was to have the honour or credit of importing the Punic bees I was engaged in.

The facts are these. I had discovered, after importing and testing every known race of bees, that the Punic race, to be found in a certain part of Tunis, were far superior to any others or to any ever suspected. For five years I was making public their remarkable and valuable qualities, and doing my utmost in vain to get them imported. At last, in 1891, I made arrangements to import them myself—going to very great risk and expense—as soon as I had got them over and had begun breeding them for sale, Messrs. Cowan and Carr said in their journals that they were acquainted with the bees in Tunis, but the so-called Punic bees did not exist there, and advised everyone to have nothing to do with them until they could give them some reliable information. As subsequent events prove, they were wilfully keeping back facts within their knowledge, as well as making statements they knew full well were false. Then the statement was made that Mr. Cowan was going to Tunis to learn all he could, and readers were advised to wait till he came back. As subsequent events also prove, he had concluded arrangements with Messrs. Roots, of America, and with one of his men, under the name of Simmins, these bees were to be "boomed," and I was to be "cut out" and left in the cold, unable to sell any I had imported. To make doubly sure of crushing my new business a letter was written to the Postmaster-General, just before Mr. Cowan went to Tunis, complaining that I was in the habit of sending live bees by post, and asking why I was allowed to break the law. This letter was signed "Thos. Wm. Cowan, Chairman of the British Bee Keepers' Association." He thus assumed your Ladyship's position for "President" and "Chairman," are interchangeable terms. It is one Mr. Cowan is for ever crediting himself with, though he has never been more than the Chairman of the Committee. If you doubt the truth of this, refer to the "British Bee Journal" for July 1st, 1897, page 256. This is the first time they let it out about Mr. Cowan going to the P.M.G.

What the Post Officials really understood was undoubtedly what Mr. Cowan intended them to understand, viz., that it was against the interests of British bee keepers that bees should be sent by post, and that the British Bee Keepers' Association were trying to stop it. I, however, referred them to dates when they had petitioned the Post Office to allow live bees to pass, and when Mr. Cowan formed part of a deputation to urge forward this matter; I also explained his object and motive in writing; so they kept a sharp look-out on live bees coming from Tunis, and sent all Mr. Cowan's back; and this was the reason he found, when he got home, that not one had been delivered.

I therefore charge that the influence of the "B.B.K.A." is being prostituted to do a honest private person a serious injury, and ruin a lawful business he was carrying on for the benefit of bee-keepers; and I charge your ladyship with being a voluntary or passive accessory. You have had ample chance to clear yourself had you wished. Now you must abide by the consequences.

You write in a fine, lofty style to the painters about "that birthright of personal liberty, under the law, inherent in every Englishman, inalienable by king or commoner." Yet you can consent to be president of a society which is practically self-elected, with no means of criticising any of their actions; which adopts every disreputable subterfuge to crush private enterprise and ruin a private person—for they did crush me, and threw my work back six years.

What is this association, pray, which you regard as such an honour to be its president? Formerly the committee were elected by means of voting papers, sent to each member; then an "improvement" was added, that subscribers of £1 and life members, who pay £5 down, should have four votes each, and only these were to be eligible for committeemen. By this means the government was transferred from the members as a whole to a clique. Voting papers were still sent out, and when received, the "clique" added theirs, so as to secure their own election. Your Ladyship was good enough to show Mr. Cowan my letter of February 12th, 1892, in which I complained of the election of the committee, and then you let your secretary, Mr. C. C. Osborne, go through the farce of counting the votes after Mr. Cowan's manager had received them. Finding this mode was getting too risky, they next got a rule made to elect the committee at the annual general meeting, but no notice is sent to members or information supplied as to who are candidates, nor are any names of candidates put on the agenda paper. These annual meetings are always held in London, at a time when very few country members can be present, but the "clique" muster up a force and carry all before them; thus they are, as I say, "practically self-elected," and your Ladyship connives at such arrangement by consenting to remain its President; while without your Ladyship's influence they would not be able to gull the public and the members as to their "disinterestedness."

The British Bee Keepers' Association was formed to teach a more humane and profitable way of keeping bees, and so benefiting the rural labourers—a very laudable object, without doubt—but how has this been carried out? Why, bees never were kept in more uncomfortable hives—hives which are a disgrace to the Royal Society for the Prevention of Cruelty to Animals, in whose rooms the meetings are held. They have wilfully spread the bee-keepers' dread disease—foul brood—all over the land; and now, having done so, they are demanding an Act of Parlia-

ment and public money to burn up every hive they say is infected, giving the owner—out of public funds—a maximum of 5s. per hive compensation, no matter what it cost; and because your Ladyship is President, they have got his Grace the Duke of Devonshire to promise to support such a barefaced piece of knavery. Besides these, they are for ever scheming to get money out of the bee-keepers in one way or another—the various dodges beat any invented by costermongers; but perhaps the most audacious of all was the British Honey Company scheme. Your Ladyship was a supporter of this, the worst scheme ever concocted, in my opinion, in connection with bees. It was a company formed by the "clique," to buy bee-keepers' honey at their own price, put on their own labels, and sell it as their own product—e.g., they were going to squeeze down the very bee-keepers they were canvassing for subscriptions to help, i.e., the were going to trade and make a profit out of their philanthropy. Not content with this, they even took powers to deal in all kinds of foreign honey. This was, they explained, when I questioned the folly of adding this, to be able to keep up their connection when they could not get British honey: so they were going to supply foreign honey under the name of the British Honey Co., with the British Isles as a trade mark. Yet your ladyship must needs be surprised at my criticising this intentional piece of "false marking"; you will find what you said in the "British Bee Journal" for May 1st, 1885, on page 145. You there say you "had not followed the subject closely," but will this exonerate you? If you choose to publicly support a scheme blindfolded you must accept all the consequences of it. That scheme, in the two years it was running, well nigh crushed bee-keeping out of existence through the low prices offered to bee-keepers for their honey; it gave bee-keeping a blow it has not recovered yet, and all to enable them to make an impossible profit. You called it "a company on sound commercial principles." Why, any lad out of a wholesale house would have laughed at such a scheme, as I told them. I could well see the evil it would do, and and the failure it would come to; and their balance-sheets proved I was correct, for it cost them 1s. 2d. to sell every shillingsworth of goods "wholesale"; with no rent to pay. If you are proud of the support—which enabled them to raise over £6000 capital—you gave that scheme, you are welcome to it. While you are referring to the passage on page 145, just read the whole report of the meeting. If your ladyship had, after consenting to be the "decoy duck," refunded each poor shareholder what he had been led to invest in it, through your influence. I might be inclined to believe in your vaunted generosity.

I did my utmost to get that scheme launched on safe commercial lines, in such a way that it would be a benefit to all; the "B.B.J." was closed against me, but the "Journal of Horticulture" was open, and in its pages you will see what I said.

I have shown in the past numbers of the "Bee Master"—copies of which were duly posted your ladyship—that no falsehood or statement is too bad for Mr. Cowan and other officials of the Association to circulate or publish; yet your ladyship still consents to preside over such a clique of "famous ones," thus branding as "sterling" all their base actions, and by the nefarious rule of claiming the president of each County Association as a vice-president of the British —whether they consent or not. They also use such names as Her Royal Highness Princess Beatrice, H.R.H. Princess Christian, his Grace the Duke of Devonshire, his Grace the Duke of Rutland, the Right Hon. the Earl of Derby, the Lord Bishop of Salisbury, Col. Lord Wm. Beresford, V.C., etc., etc. Surely your ladyship has some small spark of the fitness of things than to allow such names as these to be used as "decoys" for the purpose of inducing a confiding public to supply funds to what they call the "parent" association, the management of which is, as I have shown, composed of men who would be a disgrace anywhere.

I will omit going into particulars of how they appoint judges for shows to award prizes to previously selected parties. How the so-called "experts" are granted certificates after paying certain fees, and going through the farce of an "examination"; though ignorant of matters they certify them to be expert in. How money is got from various County Councils to be paid in fees to the nominees, called "lecturers," and the fact that up to the advent of the "Bee Master" no bee publication was open to free criticism of their actions, although the "British Bee Journal" used to boast that its pages were. All these matters, and many others, I pass over for the present to make this letter as short as possible; for, feeling secure in their position, the evils have grown from molehills into mountains.

Most people take up some form of hobby on which they spend their time and money; some go in for horses, some collecting old China, stamps, coins, etc.; whilst your ladyship, with more money to spend than most people, has gone in for the most expensive form of hobby of any, i.e., "Philanthrophy," whilst I, being a poor man, with perhaps more ambition, have been spending all my spare time and what cash I could spare to make the world more enjoyable and easier to live in than formerly. Now, whether I do this for pleasure, honour, or profit, neither you nor any association you may be president of have a right to thwart, hinder, or prevent me doing so; this is a right I possess as a Briton, inalienable by king, commoner, association, or baroness, and every society or association which conspires or uses its influence to prevent me doing so merits the scorn of all men. You may think I am doing evil work, just as I consider you are in manufacturing paupers, sycophants, and flatterers wholesale with your money, though with a pure philanthropic motive; but you have no more right to assist in trying to crush me and preventing me from working in my own way than I have to prevent you. Are you yourself as consistent? Don't try to "slate" the painters or anyone else again until you have cleared yourself of all inconsistency, and lose no time in clearing out the vile flatterers, liars, sycophants, etc., from the British Bee Keepers' Association while you have life left to do it; for death might overtake you when too late, and your fair name thus be sent down to posterity with this blot on it. You cannot shirk the responsibility by closing your eyes. You are responsible for all done in your name and under your influence. Others have been made to suffer as well as I, but perhaps I more than all, for I have combatted every injustice, and shall do so while I live, while they have either been crushed or given in. I have done my utmost to prevail on your ladyship to investigate my charges without avail; I am ready at any time and at any place to substantiate them, and if this does not induce your ladyship to see that they are investigated —if you have been an involuntary party to them—I shall next take much stronger and more unpleasant steps to get the evils I complain of removed without delay. I trust, however, your ladyship will spare me this necessity.

What I would suggest is that you have the truth of my charges investigated, and, if found correct, then call a meeting of all the vice-presidents, life members, and members to reconstruct the association on a sounder basis, or else wind it up. If, however, my charges are proved to be false, then have me prosecuted for making them.

Yours respectfully,

JOHN HEWITT.

Appendix 5

Time Line

I compiled this initially in order that I could track his movements, especially during his very active periods, and present it here in a somewhat sanitised version to provide a quick reference.

1795. John Cowan born, Edinburgh

1819. John Cowan went to St. Petersburgh, Russia

1832. 10th Nov. John Cowan m. Mary Anne Eales

1840. 2nd Jan. Thomas William b.

1842. Demetrius b.

1845 – 1854. Thomas educated at public school – St Peter and St Paul.

1854. Cowan family leave Russia for England.

1855 – 1861. T. W. C. apprentice at Rennies, engineers.

1858/9. T. W. C. registered as occasional student at School of Mines

1861. T. W. C. took charge of Kent Iron Works from his father.

1862. Kent Iron Works exhibited at London Exhibition

1861 – 1863. Various patents.

1862. First interest in bees.

1864. T. W. C. m. Fanny Matilda Michell.

1864. T. W. C. purchases Trefynant mine.

1864 – 1869 Mr & Mrs C. resided Rhyde House, Beckenham.

1864. T. W. C. commenced keeping bees.

1868/9 Kent Iron Works ceased trading.

1869. August, Alexander Henry Cowan b. Beckenham.

1869 (late) Thomas and family move to Hawthorne House, South St., Horsham.

1871. Helena Maria b. Horsham

1872. 22nd Nov. Edith Constance b.

1873. Donated box hive at Horsham Horticulture show.

1874. First article on bees.

1874, 25th Oct. Henry Michell d.

1874. Herbert Francis b.

1876. Lincs. Show Grantham, judge. Sept 11th .

1876. June. Percy John born.

1876. Purchased American dovetailed sections.

1876. Moved into Comptons Lea.

1876. April 16th. First time elected to the (new) office of Chairman.

1878. June. With Hooker and Hunter to Paris exhibition. Abbott went separately.

1878. June. Sarai Michell d.

1878. July 22nd – 24th . R. A. S. show, Kilburn, judge.

1878. 13th Aug. Lincs. Show, Stamford. Judge.

1879. 28th July. Meeting with Newman and others at Comptons Lea.

1880. 12th April. Devon and Exeter BKA AGM

1880. Aug 20th and 21st Herts BKA show, Gorhambury. Judge.

1881. 6th June. Bath and West show, Tunbridge Wells. Judge.

1881. July 15th . 1st ed. *Guide Book*.

1881. July 16th. West Kent BKA show, Chislehurst. Judge.

1881. Aug. 11th. With Peel met Postal authorities in London

1881. Aug 10th and 11th. Herts. BKA show Gorhambury. Judge

1881. Sept. 7th and 8th. Lincs. show, Louth. Spoke in evening of 7th.

1881. Sept. 14th. Surrey BKA show, London Rd. Judge.

1881. Sept. 15th, 16th, 17th. Dairy show, Islington. Judge.

1881. Dec. 2nd. 2nd ed. *Guide Book*.

1881. Dec 12th. Brighton Health Congress. Read *Honey as Food*.

1882. Declared that he did not keep bees in lofts any more, but he did have a bee house.

1882. Spent the summer in Switzerland. Met Bertrand and visited his apiary at Allevays in the Jura.

1882. September. Attended the Caledonian show.

1883. 3rd ed. *Guide Book*.

1883. 12th Jan. Devon & Exeter BKA AGM

1883. 19th Feb. Sussex BKA @ Brighton.

1883. 25th April Mrs C seriously ill.

1883. July 5th, 6th, 7th Bridgewater show.

1883. Spring/early Summer Switzerland. 10days at Zurich Ex. With Bertrand and De Layens.

1883. Aug. 7th South Kensington show. Examined.

1884. 4th & 5th eds. *Guide Book* published.

1884. Most of year spent in Switzerland, because of wife's illness. Did not return until late Oct/early Nov.

1885. 11th Feb. Appointed Trustee of BBKA

1885. Attended British Honey Co. meeting 16th April.

1885. May 18th RHS show. Examined.

1885. June 2nd. Peel dies.

1885. July 9th Bucks. BKA show, Aylesbury. Judge

1885. 23rd July. RAS show, Preston. Judge.

1885. Aug. 1st. C. becomes editor *BBJ*

1885. After 15th Aug. Travelled to Ballaigues in the Jura mountains, Ornavasso, thence to Milan for Apicultural exhibition, (24th) with Bertrand. Did not appear in Britain again this year.

1885. 6th ed. *Guide Book.*

1886. Jan. In Britain.

1886. Feb 1st. John Cowan dies.

1886. Spring. Some weeks in the Italian lakes thence to Pallanza and on to Onavasso, through Simplon pass back in Lausanne, 8th June.

1886. Mid July. RAS show Norwich, visited by Prince of Wales.

1886. 31st July Kensington ex. Read paper.

1886. 11th September. Ann Cowan dies.

1886. 6th Oct. Indian and Colonial ex. South Kensington. Met Canadian bee-keepers.

No further sightings in 1886

1886. 7th & 8th eds. *Guide Book* pub.

1887. July. Mr & Mrs sailed for New York, arrived 18th.

1887. Mid Aug. In White Mountains

1887. ~ 23rd Aug. with Cook, Michagan Agric. college

1887. Early Sept. visited Chicago, Hamilton, Medina, Niagra Falls to Toronto.

1887. 16th Sept. Canadian beekeepers presentation

1887. Philadelphia.

1887. 1st Oct left New York on SS Umbria. 'Big reception in Horsham'

1887. November. Left for 'winter residence' in Switzerland.

1888. 3rd Jan @ 8 Avenue de la Garre given as second home to the Shellys. Children there also.

1888. Feb. and March. Unwell confined to 'winter residence'.

1888. Aug. 18th. Co-operators show Chrystal Palace, Judge.

1888. Late Aug. Shropshire BKA show. Judge.

1888. 7th Sept. Lancs & Cheshire BKA show. Judge.

1888. 12th Sept Irish Ex. London. Judge.

1888. Ninth ed. *Guide Book.*

1889. Jan./Feb 'still' in Switzerland.

1889. 21st Feb. London

1889. 24th returns to Switzerland.

1889. May still in Switzerland, Lusanne. Mrs. C. very ill.

1889. 28th June. Royal show at Windsor. Presented to Queen. Came from

Switzerlad for it.

1889. Mid July. Paris ex. With Blow and Bertrand. Visited apiaries with latter.

1889. Early Aug. Cowans back in England Mrs C much improved.

1889. Aug. Cowans in Scotland then to Iceland (probably Ireland).

1889. 26th Aug Castle Douglas show. Judge.

1889. Aug. Attended ex. Kelso with Carmichael.

1889. By 17th Sept back at Comptons Lea, 'spending winter in England'.

1889. 10th ed. *Guide Book* pub.

1890. 18th Feb. Comptons Lea. Spoke at BBKA.

1890. 25th Feb. Lancs. and Cheshire BKA presentation in Liverpool on eve of amalgamation of *Record and Advisor.*

1890. May. Left for Switzerland, Lusanne, had 'been ill all winter' – see entries for 18th and 25th Feb.

1890. Late May/June toured French apiaries with Bertrand.

1890. By 21st Sept back in England, probably to 31, Belsize Gardens, Hampstead.

1891. 10/11th June. Essex show. Judge.

1891. 22nd – 26th June. Doncaster show, steward. Hewitt trouble.

1891. 9th July. Kent show. Judge.

1891. 10th – 14th July. At Bartrums, Wakes Colne, Essex.

1891. 22nd – 24th Doncaster show. Judge.

1891. 28th – 31st Highland Agric. show, Kings Park, Stirling. Judge.

1891. 7th Aug. Berwickshire BKA show, Duns. Judge with Carmichael.

1891. 11th ed. *Guide Book.*

1892. Jan/ Feb. Has La Grippe.

1892. ~ 1st April. Left for North Africa with daughter.

1892. Spring. In Switzerland at Congress – called in on way home.

1892. By mid June back in England.

1892. 30th Aug & 1st Sept. Lancs. and Cheshire show, Birkenhead. Judge. WBC.

1893. Jan ill.

1893. Late April. Left for Switzerland.

1893. ~ 6th July returned from Continent. Had spent some time in Strasburg sorting out how to analyse honey.

1893. 29th July Middlesex BKA show. Judge.

1893. 12th ed. *Guide Book.*

1893. Early June. Betrands visited.

1894. ~20th June returned from Fowey with Bertrand.

1894. 27th June Royal show at Cambridge. Attended with Bertrand.

1894. 3rd July. Show at Farningham. Present with Mr and Mrs Bertrand

1894. 4th July. At home with Bertrand and other invited bee-keepers.

1894. 18th Sept. Fowey.

1894. 13th ed. *Guide Book.*

1895. 9th May. Went to Fowey.

1895. 1st June. In Switzerland.

1895. Alexander emigrated to America.

1896. 8th May. Fowey.

1896. 9th & 10th June. Royal Cornwall Agric. show, St Ives. Judge.

1896. 22nd – 26th RAS show, Leicester. Steward.

1896. Announced that he was going to America to spend winter.

1896. 14th Nov. Mr & Mrs C. leave on SS Etruria from Liverpool.

1896. 14th ed. *Guide Book* pub.

1897. In California. Most of summer in Pacific Grove.

1897. 29th July in Oakland.

1897. Early Dec. In Oakland. Start of journey home.

1898. 10th Jan California State bk conv. in L. A

1898. 24th Feb. Loomis

1898. March. With Root in Medina.

1898. 26th May in Chicago on way to ship.

1898. 20th – 24th June. RAS show Birmingham. Judge and examiner.

1898. Late July – early Aug. Switzerland.

1898. ~ 13th Aug. returned from Switzerland.

1898. Mid Aug – early September in Scotland.

1898. 17th Oct. Announces he is going to California.

1898. Oct. Helena and Herbert drown. Mr., Mrs. And Edith leave for America.

1898. 16th November, Loomis California.

1898. 15th edition of *Guide Book* pub.

1899. 18th April. Still in Loomis.

1899. 18th Oct. Pacific Grove.

1900. Purchased Pinehurst

1900. 9th Jan. Alexander marries Mary Owen.

1900. March. 16th ed. *Guide Book* pub. Pacific Grove.

1901. 29th – 30th Aug. Read paper *The Chemistry of honey* at National bk Convention, Chicago.

1901. Late Oct. at fruit cannery in Santa Cruz mountains.

1901. Nov. 17th ed. *Guide Book* pub.

1902. Family holiday to Canadian Pacific.

1902. 21st Sept. Made hon. member of Nat. Hist. Soc of British Columbia.

1902. Met Hooker in Philadelphia.

1902. Early Nov. returned from British Columbia.

1903. ~ 20th April. Left Pacific Grove.

1903. 29th April in Chicago.

1903. 14th May arrived Liverpool on SS Saxonia.

1903. 23rd June RAS show Park Royal, London. Judge.

1903. Late July/Sept. Switzerland.

1903. 18th ed. *Guide Book.*

1904. Jan. 2nd ed. of *Honey Bee* pub.

1904. Edith ex. London.

1904. 18th Nov. Pine Grove Boscombe.

1904. 13th Dec. Pine Grove Boscombe.

1905. 11th April. Sailed Liverpool – Boston on SS Saxonia.

1905. 29th April. Chicago.

1905. May. Pacific Grove

1905. Late Aug. Left Pacific Grove.

1905. 9th Sept. Portland exposition. On way to British Columbia. Vancouver.

1905. 21st Nov. Sailed from Boston on SS ivernia.

1905. 30th Nov. Arr. Liverpool.

1906. Late. Moved to Upcott House, Nr Taunton, Somerset.

1907. July. 19th ed. *Guide Book.*

1907. 11th July. Board of Agric. London.

1908. 25th June. Franco/British Congress. Read paper.

1908. 6th Oct Dairy show, Islington. Judge.

1909. Feb. WBC dies. At all but one BBKA meetings.

1909. Sept. Grocers and kindred trades show. Judge.

1910. 20th July. Glamorgan BKA show. Judge

1910. 17th Sept. Grocers etc. Judge

1911. June. 20th ed. *Guide Book.*

1911. June. Visited Bertrand in Switzerland.

1912. Sold Pinehurst.

1912. 12th March. Attended Board of Agric & Fish.

1912. 10th Oct. Zoological Gardens, Regents Park.

1913. June. Switzerland to visit Bertrand and the Dadants.

1913. July. 21st ed. *Guide Book* pub.

1913. 1st July. Royal show at Bristol. Judge.

1918. Moved to Sutherland House, Clevedon.

Appendix 6

Supporting Cast.

A little information on some of those who only warranted a name reference in the text. The information is not intended to be exhaustive.

Ahmed Z. Abusady. Egyptian, Doctor, founder of Apis Club, editor *Bee World* 1919 – 1929. Introduced aluminium comb to Britain.

Henry Alley. American bee breeder. Developed different technique for raising queens. Somewhat inadvertently dragged into the Punic affair, subject of Cowan's attack "Humbug in the queen trade".

S. J. Baldwin. Bromley, Kent. First expert appointed by B. B. K. A. Held post for 25 years. Merchant and honey farmer. Published *Baldwin's Beekeeping Instructor* – sold with his catalogue. Died in America visiting sister. Business passed to Mrs Seadon.

Rev. E. Bartrum. Schoolmaster, priest. Ally of Cowan. Committee member.

John Charles Bee-Mason. Explorer, war photographer. Made first cine films of bees; toured halls showing them. Ended in jail having gone into local politics to highlight fraud therin. Hence not an establishment figure.

Baron A. Von Berlepsch. Invented hive, and developed it to take frames at about the same time as Woodbury, but after his countryman Dzierzon. As with most German hives it opened at the back. Credited, with Siebold with the discovery that bees mate outside the hive (~1853). Authored the *Bees and Beekeeping;* riddled with errors. Believed foul brood was the direct result of introducing moveable frame hives. Greater wealth but less intelligent than Dzierzon.

Annie Betts, Sec. Apis Club, editor *Bee World*. Prolific letter writer, author.

Frank Benton. Initially worked for Canadian bee farmer. Set up breeding apiaries in several countries and exported the different races.

Dr. Edward Bevan. Devised hive. Wrote *The Honey Bee* – two editions, third revised by Munn. Allegedly named foul brood in 1827.

Hon. and Rev Henry Bligh. Did not graduate from Oxford due to ill health. Eventually went into Church. One of the first seven that met to initiate the B. B. K. A. First Chairman of Association Committee. Invented bellows smoker. Initiator of the 'economic apiaries competition', which failed after a couple of years, as a result of the usual mismanagement. He was one of the first three B. B. K. A. examiners.

Thomas Bates Blow. Founded appliance co.which eventually became E. H. Taylor, Ltd. Went to many countries looking for races of bees to import and reporting on them. He reported on the bee farmers of America. An Association outsider often in conflict with Cowan.

Rev W. E. Burkitt. 1st class B. B. K. A. expert. Wilts County delegate to B. B. K. A.

William Carr. Commercial bee-keeper, Clayton Bridge Apiary, Newton Heath, Nr. Manchester. Wanted the B. S. frame to be 10" deep. Went to Ireland with C. N. Abbott and bee tent to preach modern bee-keeping. Fraught exercise even before they left.

Henry Chevinix. Hon. Sec. and Treasurer I. B. K. A. 1887 – 1900. Started leaflet *Notes and Hints*.

A. J. Cook. American academic entomologist. Author *Manual of the Apiary*.

Arthur J. Danyell. Ex captain H. M. 31st regiment. Author, *The Italian system of bee-keeping* which contained information how to make the Giotto hive and extractor. 1876.

J. Dennler, Nr. Stasburg, France. Schoolmaster, writer. One of Thomas's circle of European beekeepers. Visit him and vice versa.

J. G. Desborough. Stamford, Lincs. Prize winner at shows in early 1850's. Active in the initial stages of B. B. K. A. Preferred 16" x 10" frame although he was a member of the committee that decided size of British standard frame.

Dr. Johann Dzierzon. R. C. Priest. Author of *Rational system of beekeeping*. Discoverer of parthenogenesis in bees. Initially developed bar frame hive in 1838 – lacking bee space down sides of frames, initially opened at side. Lost all but 10 colonies out of 500 to foul brood in one year. First to import Ligurians into Germany (1853). In his time a real expert.

Rev. Tickner Edwardes. Prolific author – produced a book a year in his heyday, not all on bees. West Countryman, eventually became Canon. Bee editor in *Smallholder*. Never a member of the bee-keeping establishment, but considered harmless, not likely to upset the status quo.

C. J. Fox Kenworthy. Related to C. N. Abbott. Hon. Sec. B. B. K. A. June 1876 – March 1878 when he resigned. Sec. Bee and Fruit Farming Co. Auditor British Honey Co. Died age 33.

Jesse Garrat, Meopham, Kent. Typical B. B. K. A. material – active in local community. He suffered from an independent mind hence never a complete insider. Involved in the early split in the Kent County Association. Employed at local agricultural College.

Michel Girdwoyn. Exhibited *Anatomie at Physiologie d l'Abeilles* at the Paris exhibition, 1876. Cheshire acquired copy and appeared to claim ownership, but others also had them. B. B. K. A. purchased them from Cheshire and published them as coloured wall charts. Used by several authors without acknowledgement.

R. R. Godfrey, Grantham. Hon. Sec. Lincs B. B. K. A. Abbott supporter.

R. A. Grimshaw. Leeds, Yorkshire. Nail manufacturer. Poor education but contributed to the *Garden* (not bees). Read several papers at conversaziones on bee physiology. Invented 'Apifuge', deterrent to stings. Staff member of *B. B. J.* (wrote regular column), from 1887, thus Cowanite.

James Heddon. American bee farmer. Controversial figure. Initially used 8 frame Langstroth hive. Author *Success in Bee Culture*. Devised hive – two boxes per brood chamber, closed end frames; did not find favour with Cowan who also claimed that Heddon was prosecuted for adulterating honey – I found no evidence.

Otto Hehner. B. B. K. A. annalist of honey and thus friend of Cowan but removed from Christmas card list when he disagreed with Cowan over whether pure wax could be white; Cowan claimed that it was impossible.

George Henderson. Ealing Middlesex. Worked in printing business, Greek and Latin scholar. Sub editor of *B. B. J.*, 1873 – 1889. B. B. K. A. Council member and librarian.

Hoge. Initially agent in Britain for Thurber an American honey dealer. British bee-keeping establishment claimed his honey was adulterated. Claimed to be purveyor of honey to Queen. Set up own company which failed and he returned to America.

J. M. Hooker. Architect, home counties resident. Designed Alexander hive. Member of committee that decided size of British standard frame. Claimed to be inventor of uncapping knife. Wrote hardly known *Hookers Guide to Successful Bee-keeping* a *B. B. J.* booklet. Later life spent in America. Establishment figure, and their voice in the *Journal of Horticulture*.

John Huckle. Kings Langley, Hertfordshire. Never a bee-keeper or wealthy. Active local figure. Office manager of *B. B. J.* and paid secretary (£25 then £50 p. a.), of B. B. K. A. Just did what first Peel and then Cowan told him. Initially known to Peel through cricket.

Dr. C. R. Killick, Somerset. V. P. B. B. K. A. Major activist in move to reorganise the B. B. K. A. Expelled in 1928. Founded National B. K. A. same year, ultimately becoming its President.

G. De Layens. Devised the hive that was effectively the national hive of France. Studied foul brood, reporting on his experiments in great detail, unlike his friend Cowan who was apt to ask readers to just believe him.

R. O. B. Manley. Bee farmer from age forty. Author *Beekeeping in Britain* etc. etc. Opponent of legislation as proposed by B. B. K. A.; proponent of larger frame, thus anti BS frame size: critic of the cost of the *B. B. J.* Passed 3rd class expert, examined by Herrod –strongly critical, did not take further exams. Edited *Practical Bee Guide* after Digges death.

Piers Edgecombe Martin. Stockbridge, Hampshire. Master mariner. Devised 'sailors' hive, which he used on his be farm, thought by many to be the forerunner of the present standard National hive.

W. McEvoy. Canadian. Credited with introducing the successful method of foul brood treatment in his country.

Alfred (son of George) Neighbour. Merchant, Regent St. London. High Holborn, London. First to import Ligurians and Carniolans, friend of T. W. Woodbury. Sent honey-bees to Australia and bumble bees to New Zealand. Wrote *The Apiary,* an extended catalogue. Member of committee to decide on frame size for British standard. Large library sold to H. J. O. Walker.

T. G. Newman. Born Somerset, England. Magazine publisher and editor. Proprietor and editor *American Bee Journal* 1873 – 1892. President National Beekeepers Association; Manager National Beekeepers Union. Friend of Cowan, visited England several times.

W. J. Pettitt, Dover. Very early hive manufacturer – generally very elaborate efforts. In 1834 made a hive in Mahogany lined with Cedar. Double walled?

E. L. Pratt. American queen breeder. Introduced wooden cell cup and spiral wire cell protector. Conducted queen breeding part of *Apiculturist,* Then started *Queen Breeders Journal* - very short lived. Closely associated with Alley, at least in Cowan's mind, and the attack on the pair showed Cowan's lack of understanding of beekeeping in America despite having travelled and visited many beekeepers there.

William Raitt. Blaigowrie, Scotland. Beekeeping came second in his life to religion. Schoolmaster, equipment dealer, bee farmer although not very commercially orientated. Much development on hives, foundation, honey press etc. Joint editor *The Beekeepers Record,* editor of bee section in *Journal of Horticulture* at some stage. Friend of Cowan from early days.

Rev. G Raynor. Cambridge graduate. Allegedly introduced carbolic acid for subduing. Invented hive, queen introduction cage, bottle feeder, extractor, divisional section rack. One of first to examine for expert, member committee for British standard frame. Early correspondent to *Journal of Horticulture* and *Field.* Resident Australia mid life. Author *Useful Hints* in B. B. J. Cowanite.

Abbe Della Rocca. French and Greek islands. Invented hive (1790) based on Greek basket hive, including self spacing bars. First with champhered top bar. Author. Many strange bee beliefs.

Alfred Rushbridge. Chichester. Published *Book for Beekeepers* and *Beekeeping plain and Practical and how to make it pay*. Nothing more than glorified catalogues for his business.

A. G. Schirach. Saxon clergyman. Cowan and therefore W. Herrod-Hempsall believed Schirach to be the first to recognise and write about foul brood in 1769. He was certainly the first to demonstrate that queens could be raised from worker eggs up to three days old simply by the food they were given. But he also believed that sexual intercourse did not take place in honey bees and that the queen was self-fertilising. Germany's Huber.

Pastor Paul Schonfeld. Germany. Researched the anatomy of honeybees, and the cause of foul brood. Cowan claimed that he had discovered that which Cheshire claimed to be his find, ten years earlier. His first hypothesis appeared in 1874, and was translated into English by J. S. Wood. The discovery was earlier by Dr Preuss.

Rev. F. T. Scott, appointments in Kent, graduate and tutor Oxford. Early importer of Ligurians. Developed hive. Involved from the very beginning of B. B. K. A. Vice President. Member of committee to determine size of frame for British Standard.

Samuel Simmins. Sussex. Country newspaper proprietor. Author *A Modern Bee Farm;* several leaflets – advertising material really. Published *Bee Chat* a magazine – short lived. Attempted every possible form of beekeeping business – honey production, queen rearing and hive manufacture, failed every time. Devised divisional, Conqueror, Union and Economic hives; none successful. Manley described his theories and methods as 'peculiar'.

Captain F. Sitwell, Wooler & Berwick, Northumberland. Council member 1910. Supporter of legislation, major figure in Northumberland beekeeping.

William Thompson, Auchinraith, Blantyre, Scotland. Aka 'The Lanarkshire Beekeeper'. Regular correspondent and one time editor of bee section in the *Journal of Horticulture.* Laid claim to nearly every beekeeping invention, either for himself or another Scot generally without evidence. Big advocate of the Stewerton hive.

E. D. Till. Eynsford, Kent. Inherited wealth and spent it on local causes. He was a B. B. K. A. Council member 1893 – 1909, having commenced beekeeping only two years earlier. Strong advocate of legislation and one of the few who appreciated the need for statistics to support the case and did his best to provide them for Kent. However he was incapable of stopping the collapse of the Kent B. B. K. A. when a large proportion became disillusioned by the central Association. Very much a Cowan supporter.

Timberlake. A. & H. Timberlake were honey and wax merchants in East London. They attempted to take a large stake (debentures), in and control of the British Honey Company – rebuffed. They held £500 shares in the follow-up British and Irish Honey Co. managed by Alfred Timberlake, which also failed but the Timberlakes would not have lost money. A. Timberlake also accompanied Carr and Abbott on the Irish expedition, but is never mentioned.

C. T. Tite. Somerset. Long term, close friend of Cowan. First to broach subject of County Associations and was active in his native County.

C. N. White, Somersham then St Neots, England. Schoolmaster. Hon. Sec. Hunts. B. K. A. Regular contributor to *B. B. J.* and *Field* using pseud. 'Ivo', until 1894/5 when he moved from being part of the establishment to outcast. 1st class expert and held foul brood extension. Author *Pleasurable Beekeeping* and *Bees, a Practical handbook*.

William Woodley. Contributed regular column to the *B. B. J.* but unlike others he appeared to have a mind of his own, occasionally expressing opinions 'off message'. He operated up to 200 colonies and like many Cowan associates was active in his local community.

Acknowledgements

A single line entry does not really do justice to the effort put in, or the lengths gone to. This small list represents many hours work and is testament to the generosity that I have experienced mostly in compiling this work.

Aberg, Sally, & Beales, Don. The Heritage Society of Pacific Grove, California.

Archivist, Brighton College.

Barrett, Anne. Archivist, Imperial College.

Benny, Anne. Budock Church office.

Brake, Prof. Laural Brake, Birkbeck, University of London.

Burgess, Ken. Bishops Hull.

Eagle, Phillip. The British Library.

Harrison, Carol, Robinson Library, Newcastle upon Tyne University.

Ivamy, Reg. Headstone photo.

Kelly, Roger. Genealogy of the papermaking Cowans of Scotland.

Librarians at Newcastle upon Tyne central library.

Littlewood, Seb. Beamish Museum.

Horsham Museum.

Michell, Glenn, America. Genealogy.

Mills, Mary, and members of the Greenwich Industrial Archaeology Society.

Morgan, Carol, Institution of Civil Engineers.

The Patent Office.

Ryan, Jim. Digges and the I. B. J.

I wrote to every University in Britain that could award a doctorate pre 1871, seeking confirmation of T. W. Cowan's claimed Ph.D. I carried out a similar exercise, mailing every awarding Institution in both Britain and the U. S. A. asking if they had conferred the D. Sc. that my subject believed he had. Amazingly, I received a return of 87%, all negative. I thank all that responded.

References.

American Bee Journal. 1861 – 1926.

Beekeepers Record. 1882 – 1926.

The Bee Master. 1897 – 1898.

Bee World. 1919 – 1935.

British Bee Journal. **1874 - 1926.**

English Mechanic and World of Science. 1965 – 1925.

Gleanings in Bee Culture. 1873 – 1926.

Journal of Horticulture. 1865 – 1926

Irish Bee Journal. 1901 – 1933.

The Times. 1875 – 1926.

Anon. *Instruction in Bee-keeping for the use of Irish Bee-keepers.* Dept. of Agric. & Technical Instruction for Ireland. 1905

Anon. *Structure of the Bee.* Gresham. 1903.

Barnes, Eleanor C., *Alfred Yarrow, his Life and his Work.* Edward Arnold & Co. 1923.

Cheshire, Frank R. *Practiacl Bee-keeping.* The 'Bazaar' Office. 1873.

Cheshire, Frank R. *Bees and Bee-keeping.* Upcott Gill. 1866

Cowan, T. W. *The British Bee-keeper's Guide Book.* Eds. 1-25, 1881 - 1924.

Cowan, T. W. *British Bee-keepers Practical Note Book.* Huckle. 1884.

Cowan, T. W. *The Honey-bee.* Houlston. 1890, 1904.

Cowan, T. W. *British Bee-keepers' Association Jubilee.* British Bee Journal. 1928.

Cowan, T. W. *Waxcraft.* Samson, Low, Marston. 1908.

Cowan, T. W. *The Queen Bee.* British Bee Journal. 1910

Charles, David. *Somerset Beekeepers and Beekeeping Associations.* Somerset Beekeepers' Association. 2005.

Digges, J. G. *The Irish Bee Guide.* Irish Bee Journal. 1904.

Digges, J. G. *The Practical Bee Guide.* Irish Bee Journal etc.1910 – 1950.

Fraser, H. Malcolm. *Beekeeping in Antiquity.* University of London Press. 1931.

Fraser, H. Malcolm. *History of Bee-keeping in Britain.* Bee Research Association. 1958.

Graham-Smith, G. S. et al. *Report on the Isle of Wight Bee Disease and supplements.* H. M. S. O. 1912 on.

Herrod, William, *Producing, Preparing, Exhibiting and judging bee Produce.* British Bee Journal. 1912.

Herrod-Hempsall, William. *Bee-keeping New and Old.* British Bee Journal.

1930, 1937.

Herrod-Hempsall, William. *The Bee-keeper's Guide to the Management of Bees in Moveable Comb Hives.* British Bee Journal. 1938.

Harding et al. *British Bee Books a Bibliography 1500 – 1976.* International Bee Research Association, 1979.

Harker, Leonard S. *Blazing the Trail.* G. W. Daniel Co. Ltd. 1938.

Langstroth, L. L. *The Hive and the Honey-Bee.* Hopkins, Bridgman & Co. 1853.

Lightning, R. H. *Charles Nash Abbott.* Ealing Local History Society. 1968.

Manley, R. O. B. *Bee-keeping in Britain.* Faber & Faber, 1948.

Morse, Roger A & Hooper, Ted. *The Illustrated Encyclopedia of Beekeeping.* Blandford Press1985.

Naile, Florence. *The Life of Langstroth.* Cornell University Press. 1942.

Neale, Kenneth, *Victorian Horsham.* Phillimore. 1975.

Pellett, Frank Chapman, *History of American Beekeeping.* Collegiate Press Inc. 1938.

Phillips, E. F. and White, G. F. *Historical Notes on the Causes of Bee Diseases.* U. S. Department of Agriculture. 1912.

Root, A. I. *An Eyewitness Account of Early American Beekeeping.* The A. I. Root Co. 1984.

Root. *ABC and XYZ of Bee Culture.* 1877 – 1926.

Simmins, S. *A Modern Bee Farm.* T. Pettitt & Co. 1887.

The Times Bee-master. *Bee-keeping.* Sampson Low, Son, and Marston. 1864.

Various. *Modern Bee-keeping.* Longman Green, 1880.

Walker, H. J. O., *Catalogue of Bee Books.* Author, 1929.

Watson, James K. *Bee-Keeping in Ireland.* Glendale. 1981.

Weedon, Alexis. *British book trade archives 1830 – 1939.*

Weedon, Alexis. *Victorian Publishing.* 2003.

Wilson, A. N. *The Victorians.* Hutchinson. 2002.

* 9 7 8 1 9 0 4 8 4 6 7 5 8 *